S0-BPJ-364

Future Codes

Essays in Advanced Computer Technology and the Law

The author is grateful for the opportunity these publications provided for the first publication of the following essays and for their permission, as noted by acknowledgment of copyright, to reprint the items in this collection.

Chapter 2, "Molten Media and the Infiltration of the Law," was first published in a slightly different version (electronically) at 3 *Leonardo Electronic Almanac* (June 1995).

Chapter 4, "The Uneasy Treaty of Technology and Law," AI Expert, was first published in a slightly different version at *Virtual Reality Special Report* 33. © 1994 Miller Freeman Emerging Technologies Group.

Chapter 5, "Data Morphing: Ownership, Copyright, and Creation," was first published in a slightly different version at 27 *Leonardo* 117, No. 2 (1994). © 1994 MIT Press and *Leonardo.*

Chapter 6, "Copyright Issues on the Net," was first published in a slightly different version at 22 *San Francisco Attorney* 25 (Oct./Nov. 1996).

Chapter 8, "Alters," was first published at *Wired* 114 (Nov. 1993). © 1993 *Wired.*

Chapter 10, "The Encrypted Self: Fleshing Out the Rights of Electronic Personalities," was first published in a slightly different version at 13 *John Marshal Journal of Computer & Information Law* 1 (Oct. 1994). This was an outgrowth of an earlier note published as "The Electronic Persona: A New Legal Entity," *Virtual Reality World* 37 (Jan./Feb. 1994).

Chapter 11, "Liability for Distributed Artificial Intelligences," was first published in a slightly different version at 11 *Berkeley Technology Law Journal* 147–204 (1996) (a continuation of *High Technology Law Journal*).

Chapter 12, "Recombinant Culture: Crime in the Digital Network," was originally delivered at DefCon II (Las Vegas, July 1994), published electronically, and also reprinted, DeVry Institute of Technology, *Readings in Social Issues in Technology,* 2nd ed., 1996.

Chapter 13, "Encryption and Export Laws: The Algorithm as Nuclear Weapon," was first published in a slightly different version at *Software Publisher* (Nov./Dec. 1994).

Chapter 15, "Implementing the First Amendment in Cyberspace," was first published in a slightly different version at *MultiMedia Review* (Summer 1993). © 1993 Mecklermedia Corporation.

Chapter 16, "The Electronic Word: Democracy, Technology, and the Arts," Review, was first published in a slightly different version at 16 *Hastings Comm/Ent Law Journal* 501 (Spring 1994). © 1994 by University of California, Hastings College of the Law, reprinted from *Hastings Communications and Entertainment Law Journal* (Comm/ENT), Vol. 16, No.3, pp. 501–516, by permission. (Reprinted here with additional material.)

For a complete listing of the *Artech House Telecommunications Library,*
turn to the back of this book.

Future Codes

Essays in Advanced Computer Technology and the Law

Curtis E. A. Karnow

Artech House
Boston • London

Library of Congress Cataloging-in-Publication Data
Karnow, Curtis E. A.
Future codes : essays in advanced computer technology and the law / Curtis E. A. Karnow
p. cm.
Includes bibliographical references and index.
ISBN 0-89006-942-5 (alk. paper)
1. Computers—Law and legislation—United States. 2. Technology and law. I. Title.
KF390.5.C6K37 1997
343.7309'99—dc21 97-14544
CIP

British Library Cataloguing in Publication Data
Karnow, Curtis E. A.
Future codes : essays in advanced computer technology and the law
1. Computers—Law and legislation 2. Electronic data interchange—Law and legislation
I. Title
343'.078'004

ISBN 0-89006-942-5

Cover design by Jennifer Makower. Background image by Annette Karnow.

International Standard Book Number: 0-89006-942-5
Library of Congress Catalog Card Number: 97-14544

10 9 8 7 6 5 4 3 2 1

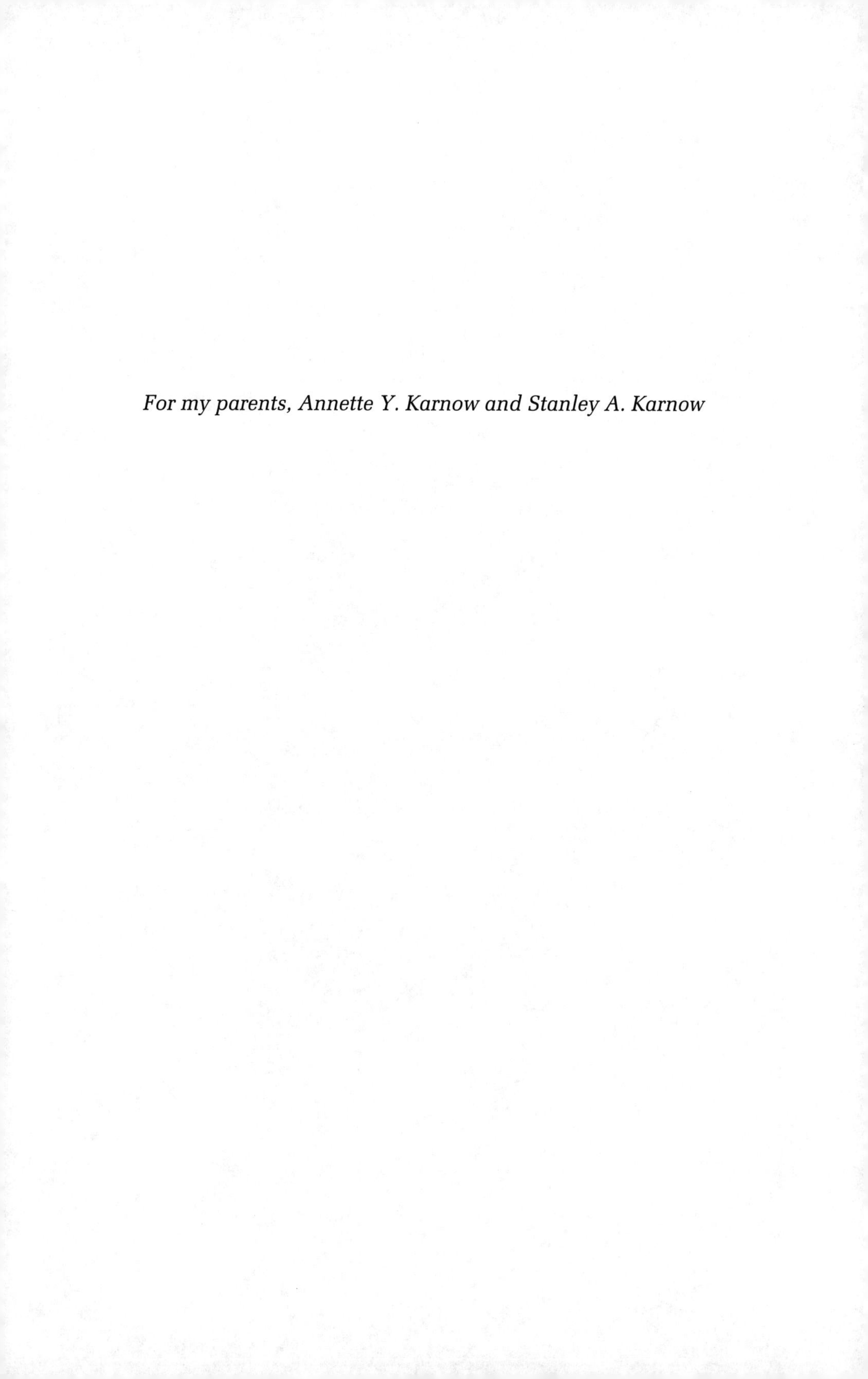

For my parents, Annette Y. Karnow and Stanley A. Karnow

Computers...cause extreme instability. They flatten corporate hierarchies. They are viral. They are portable. They undermine natural boundaries. They die out fast. They are disposable. They change so rapidly it's impossible for the whole population to keep up with them.

—Bruce Sterling, "ACM Speakers Corner"

The wheels of justice grind exceedingly slowly, but they grind exceedingly fine.

—Anon.

Contents

Part I Introduction

Everything that can be invented has been invented.

—Charles H. Duell,
Commissioner of the United States Patent Office, 1899,
urging that his office be abolished

LAW AND TECHNOLOGY

These are essays in an old sense, *essais* written in order to test my response to the subject [1]. They are forays, explorations of a new country where the old codes of law do not always mesh well with the new codes of advanced computing technology.

These must be experiments, on the fly; for technological events are moving too quickly to allow for assured doctrine. (As I edited this collection, I found myself changing future to the past tense, replacing speculation with the recitation of now-historical fact.) So these dispatches are not always consistent, and they approach some concepts, outlined in this introduction, from a variety of viewpoints and with a variety of rhetorical devices. I have used short bits of fiction, flowery prose, and strict academic exposition to cut into the new phenomena from different angles. Some of the positions I take are wrong; or they will turn out to be wrong.

These essays were written and given as talks over a period of a few years as I developed an understanding of the reciprocal effects of (1) the legal system and (2) the business of my clients who were in or affected by computer and Internet technology. As I counsel clients in business transactions and represent them in court, I have seen the fields of law and technology thrown against each other, necessarily but often antagonistically. Like enemies cast in a lifeboat, the realms of technology and law are forced to face each other, despite their differences in pace and intention.

In this quiet maelstrom, there is room for a mediator. I hope these essays will serve to introduce the legal issues to the purveyors of technology, and to welcome lawyers and students of the law to the possibilities and constraints of computing technology.

ACCELERATED TECHNOLOGY

My own interest in computers goes back to the late 1960s, when I adjusted voltages on analog computers and explored the BASIC computer language on a terminal linked into the Dartmouth College timesharing system. Since then, I have risen to the unimaginable glory of writing a tiny airplane simulator for the Tandy 100 laptop computer, counting every byte of memory. I have investigated the secrets of the CP/M operating system and finally surrendered to the luxuries of fat memory and fast computer applications written by large corporations in Seattle. I have reveled in the augmented power of the machine and have always found it, just below the surface, cryptic.

From the 1960s, which marked the availability of at least shared computing power to the general population, to the early 1980s, which brought complete, integrated computers to the home and office desk top, through to today's widely distributed computing web, we have witnessed increasingly accelerated advances in computing power, application, and availability. Concomitantly, we have witnessed the increasing insinuation of computers in our lives.[1] As a result, issues that used to have nothing to do with machines are now machine issues, and legal problems that had nothing to do with computer law are now the province of computer lawyers, requiring their special expertise in technology and the related legal issues of copyright, trademark, and so on. As the computer industry has moved from the garage to the *Fortune* 100, its governors moved from jeans to suits—and its disputes moved from the sand lot to the federal courtroom.

And as competition heated, the race to market accelerated. So we have seen increasingly short product cycles, increasingly rapid mutations in operating systems, and versions of software distributed at increasingly fast rates. With Internet distribution, computers can now update them-

1. For example, in 1984, 25% of Americans worked with computers; in 1996, the number was up to 50% [2]. A plethora of examples of this insinuation are found in the essays in this collection.

selves with new software coming out every few months. Operating system and chip production cycles are far shorter as well.

One measure of the overall acceleration of computing "progress" is the changing cost of information processing; that is, the dollar cost per computer instruction per second (IPS). On a logarithmic scale, where the year 1975 = 100, *The Economist* reports these numbers [3]:

Year	*Typical Machine*	*Log Scale Cost (Dollars per IPS)*
1975	IBM mainframe	100
1978	Digital VAX	10
1986	Sun Microsystems	1
1994	Pentium PC	.01

Remarkably, 70% of the computer industry's revenues come from products that did not exist two years ago [4].

LEGAL DELIBERATIONS: REASONING FROM THE PAST

The more corrupt the state, the more numerous the laws.

—Tacitus

The traditional legal system, by contrast, is as deliberate as molasses. It is a complex system involving all branches of government. Regulations, such as those administered by the Federal Trade Commission, the Securities Exchange Commission, state insurance watchdog agencies, and hundreds of other agencies, are enacted by the Executive branch. Laws on criminal and civil liability are enacted by the legislatures. The Judicial branch interprets and applies those rules, as well as developing and applying the rules of the "common law," which have grown up more or less solely within the province of the courts. Federal law and the laws of the various states mix and blend with each other, with the courts as the arbiters of which law will govern in a given circumstance. State and federal courts may vie for dominance over certain issues, and trial courts make an uneasy truce with their appellate court supervisors. Legislatures, both state and federal, act to limit and expand court rulings, and the courts in turn construe and apply, in sometimes unpredictable ways, the enactments of the legislatures.

These processes, considered elements in the development of legal doctrine over time, are ponderous.

Consider the life of a lawsuit. In the California state courts, most cases get to trial between two and four years after the initial complaint is filed. The losing party may appeal the judgment to the appellate court, roughly a two-year process. The losing party then has a chance at review by the Supreme Court, which may take anywhere from a few months if the High Court rejects the appeal to up to another few years if the case is taken for plenary Supreme Court review. Occasionally, the net result of the appellate review is to send the case back to the trial court for another trial, and the process starts all over again.

The legal process can be unwieldy and costly for those caught up in its web. For that reason, most cases do not in fact go through the utterly painful process outlined just above: that sort of full-court press is reserved for well-funded cases, wealthy clients, and cases likely to have substantial precedential value (i.e., where the result in one case sets the tone and perhaps result for subsequent matters). Most cases—about 95% in the California state courts—settle before trial.

But as we consider the development of the law as a source of rules, with predictive and planning value for the rest of the population, it is not the settled cases that are of interest. Lawyers and judges do not, after the fact, pay much attention to those. Instead, those in the legal business look at the published opinions of the appellate courts, including the supreme courts of the various states and of the federal government, as they try to tease out the meaning and effect of the law. It is in this context that the deliberate process of the legal system confounds those who would apply its principles.

Consider, for example, an attorney asked to evaluate the odds that a certain software copyright suit may succeed, or to provide language in a software licensing contract that will ensure her client retains rights. Or perhaps she is asked whether there are trademark and trade dress rights associated with the appearance—the look and feel—of a graphical web page. The copyright and trademark statutes, last amended some years ago, will be of moderate help, but like all statutes these contain words of uncertain meaning and uncertain scope. How could it be otherwise? A statute that contemplated every eventuality would be larger than the universe itself. And anyway, lawyers delight in ambiguity, hypothetical questions, the agnosticism of meaning: We know that most every phrase may be disputed someday, if it is in someone's interest to dispute it.

So the lawyer, or judge with a case to decide, turns to cases interpreting and applying the statute. This will at least narrow the scope of contention. These cases, say, are rather recent: Fortunately, one finds a case decided a year ago. That case lay in the court of appeal for perhaps two years, after (at best) two years in the trial court. The trial, in turn, probably addressed law, technology, and parties' actions as of about a year prior to the instigation of litigation. Thus, in an important way, the lessons of the case are about five years old—and that is for a relatively recent case. Most cases deal with older technologies and products and the actions of people that occurred farther back in time than that. It is in this context that we recall the two-year period alluded to earlier, during which 70% of the computer products now being sold were developed.

Of course, the law is *always* "behind the curve," in a sense, regardless of the industry involved and regardless of the area of law. Lawyers, like everyone else, reason by analogy from old precedent and old statutes, some it very old indeed, such as the Constitution and the later Bill of Rights. Every case in some way presents new facts, new products, new and different events to evaluate in the variable light of relatively venerable legal principles. And so judges and lawyers make analogies, suggesting that the similarities between cases are important—or irrelevant, depending on how favorable or unfavorable the case appears to be. A breach of contract involving delivery of steel may be treated very much like the delivery of cardboard containers. Theft of customer lists may be treated much like the abstraction of a product formulas, and so on.

But this "reasoning from analogy" to which lawyers are trained presupposes a lurking metaphor and overarching imagery, shared assumptions as between the old and the new contexts. Absent such commonalities, we will have no assurances that our predictions about the law—about what a judge or jury will do—are correct. We will not know which aspect of the older cases may be pertinent to the new scenario. Absent a persistent, underlying metaphor, predictive ability diminishes, and it becomes increasingly difficult to counsel clients.

Uncertainty makes cases difficult to resolve and indeed changes the bases upon which they are resolved. For when the legal rules are uncertain, litigation is increasingly resolved not on the basis of the merits of the case—because no one has a firm handle on the merits—but on the basis of avoiding the transactional costs of litigation (its costs, the attorneys' fees, and the expense of having company executives tied up in hearings and depositions). As litigation increasingly is resolved on the basis of expediency and economic staying power, a culture of disillusion sets in, and the law as a moral force, as an arbiter among competing values, loses its power. Then the law cannot credibly manage and perpetuate overriding metaphors. It seems less the forum of justice; we pay less respect to the legal merits—and that is the vicious circle.

It is one of the themes of this collection that, certainly in the area of advanced technology, the legal system is increasingly unable to rely on shared assumptions, and there is increasing difficulty in reasoning by analogy to new facts. Traditional legal metaphors, which have always guided lawyers and judges as they articulated and applied the law, enjoy a diminished authority for a variety of reasons. Four of those reasons are outlined in the next section.

UNDERMINING THE POWER OF LEGAL ANALOGIES

Evisceration of Old Products, Services, and Distribution Channels

First, and perhaps most importantly, advanced computer technology undermines the assumptions of older categories. For example, interactive networked hyperlinked media eviscerates the idea of authorship, and with it one of the fundamental concepts of U.S. copyright law. These products may be made by a variety of anonymous persons, and indeed partially or wholly written by software. Another example is trademark law, which depends closely on the notion of specified markets and product classes and categories purchased at distinctive locations under circumstances that differ according to the product class. Cars are examined and purchased at car stores, wine is sold next to other wines and alcoholic beverages, and magazines are sold at magazine racks. Trademark law asks whether consumers may be confused between products in specific contexts. But on the Internet, virtually an infinite variety of products and services compete "next" to each other. A further example is the existence of local-area networks, intranets, and the Internet, all of which contribute to the evolution of new relationships between those who pay for work and those who get paid. We have telecommuters, virtual corporations, outsourcing, closely knit strategic relationships between physically separate companies, and a variety of other unique relationships.[2] Because much traditional law (copyright ownership, certain patent rights, privacy issues, and so on) depends on whether a relationship is or is not that of

2. The work of Nobel Laureate Ronald Coase (1937) suggests workers organize in firms to minimize transaction costs. The size of the firm is determined by the relative cost of buying services from outside the firm and the overhead cost of providing them inhouse. Information technology—the Internet, facsimile, and so on—reduces the costs of buying services from the outside. Peter Huber has noted that, therefore, the traditional rationale for firms disintegrates as it becomes increasingly rational to purchase goods services from the "outside" of the firm. Thus, traditional employees are replaced by "outsiders" linked by networks. I would add that, concomitantly, the distinction between "inside" and "outside" becomes increasingly fuzzy, and so an increasingly inappropriate basis on which legal analyses rest.

employer/employee, advancing technology eviscerates the presumptions of that law.

Shifting Technology

Second, and closely related to the first point, is that advanced computer technology conflates distinctions that made much sense under older regimes and which informed law that grew up in the older regimes. New technology eviscerates the distinctions between public and private, the telephone and mail, the written and spoken word, broadcasting and point-to-point communications, and between the publication, consumption, and distribution of information. As a result, there is disagreement on whether communications between computers is like public exchanges in a town square, or like a bookstore with a thousand books, or like a newspaper. But doctrines of free speech and other rules depend on which of these metaphorical structures is thought to prevail. Indeed, computer technology is advancing so fast that older distinctions in the computer industry itself are quickly losing their force, such as those between object and source code, client and server, program and data, and passive web distribution, broadcasting, and pointcasting.

Conflation of Human, Natural, and Technological Agents

Third, increased automation, with a concomitant reduction in the role of effective human oversight, creates difficulties in the assignment of liability or legal blame. Modern automated systems control, or assist in the control of, large networks such as the telephone system and the power grid, as well as classic computer networks that provide the communications backbone for large distributed corporations and universities. Automated systems control complex aircraft and play an important role in medical diagnosis and treatment, among many other functions. Failure occurs in these complex systems as a function of the interplay of multi-component software, the amalgamation of which cannot always be foreseen by their human architects and software authors. Indeed, it is commonplace to observe that no single human can possibly understand the range of possibilities inherent in common complex software products, such as operating systems like Windows 95. Yet, because we know that at some level, by necessity, humans are entirely responsible for the computer environment in which the failure occurred, we have difficulty relegating the failure to the forces of nature, acts of God, or other default sources that are shorthand for the lack of human legal liability. In brief, we no longer are able to invoke the convenient dichotomy that divides the causes of injury and failure between human and natural forces.

The Loss of Shared Assumptions

The legal system is inhibited in its use of traditional metaphors and analogies for a fourth reason. The pace of technological change is not only rapid, it is, more importantly, highly uneven. Whereas we may have a relatively coherent and congruent set of assumptions about the way the physical world works, we do not have that common basis in the fabricated world of the computer, in what we might call the *electroverse.* Slowly developing technologies, by definition, have more time to percolate through the culture than do the quick ones. And when the pace of change is especially fast, segments of the culture will simply not assimilate. The culture will not gracefully absorb changes and mutate. Rather, it fractures and balkanizes—like putty contorted too quickly for its structure to compensate. The existence of new technologies, such as the Internet or Java programming, is simply not necessary for the work and play of many people. Not everyone needs a semi-intelligent autonomous software agent to gather up information on the World Wide Web. For those just discovering cellular phones or VCR programming, the ability to have a computer automatically update its own software may not be either interesting or understandable.

Concomitantly, there are few generally shared understandings about the capabilities of computer technologies. Not everyone knows that cellular phone calls are easily intercepted; not everyone knows what a Tempest attack is (the remote detection of the faint electromagnetic signals given off by the screen and keyboard of a computer); not everyone is aware that deleted files can often be recovered. Few have an understanding of the mechanism of a typical Internet communication, or how data packets move through the Internet backbone, or where the data "is" that they view when they click on a web page icon. Many users of local-area networks do not know that others can view their "home" directory.

Now the law delights in tests that ask whether a defendant took "reasonable precautions under all the circumstances" to ensure something was done or avoided. In the context of technologies (such as the car and telephone) that slowly develop as they percolate through the consumer base, courts can correctly assume a widely shared consensus on the capabilities of technology. The courts can therefore maintain their credibility as they judge our actions by the putative norm. There is a general consensus on the cause-and-effect relationships between the actions of a driver and the consequent behavior of a car. In the physical world, we often secure a consensus on what should be treated as private and what is up for public grabs: for example, most agree that opening an envelope is a violation of one's privacy rights, as would be opening a locked desk drawer, or crossing onto private property (we understand what "private property" means in the physical world) to look in a bedroom window. But because users of computer technology have widely different levels of knowledge about the technology, there is no general consensus on whether an employer should be able to read employees' encrypted mail, which was created and mailed on company time and on company machines. Users do not have common views on the risks posed by clicking on Java or ActiveX applet icons on a web page, primarily because most users do not know what those are or what they may do to one's local hard drive.

This lack of common understanding about the capabilities of the technology is not equivalent to ignorance about the inner workings of, say, an automobile. The mechanisms hidden under the hood of the car do not seriously affect important user expectations: We can afford to treat the car as a black box, to pay attention to a few gauges and let it go at that. By contrast, user understandings and expectations of the capabilities of local area networks, for example, will directly affect users' beliefs on the privacy and security of their data, whether a user is truly anonymous, and so on. Knowing how data packets move from machine to machine, and familiarity with data packet "sniffers," will inform one's beliefs on the integrity of data and the potential for economic espionage. The law uses those sorts of beliefs to judge whether one's "reasonable expectations" of privacy have been breached in deciding privacy suits; the law uses those sorts of beliefs to decide whether one should be liable for the unintended release of a trade secret.

What, then, is it about computers that sets them apart from other consumer products? (For it is most obviously in the *consumer context* that we see the differential rate of knowledge diffusion.) There is a clue in the way we tend to think of computers as potentially "intelligent" in a way that one never treats cars, telephones, or airplanes.[3] This is because phones, cars, and washing machines are all defined by one technological purpose—transportation, voice communications, clothes cleaning, and so on. Computers, as Alan Turing formalized the notion, are by definition *universal* computing machines: They are machines which, given time, can accomplish any task that is essentially computable—and that is an infinite number of tasks. Humans seem to share this potential. Hence the intuition that computers are, or can be, intelligent.

---

3. To the extent one *is* tempted to treat cars and planes as intelligent, it is of course because of the computers on board. And there are more and more computers in vehicles: "'Today, if you change a line of code, you're looking at the potential for some major problems. Hardware is very predictable, very repeatable. Software is in much more of a transient state.' The volume of code is exploding as processors proliferate behind the dashboard and under the hood. The typical auto has 10 to 15 processors; high-end cars can have as many as 80.... 'An engine controller can have 100,000 lines of code [according to a Bosch VP]'" [5].

To be sure, telephones, planes, and cars can be complex, too. Some of these are extremely complex but they are all (aside from their onboard computers) *bound* in their complexity. These machines are bound to the purpose to which each is devoted. It is for this reason that users can react to each of these machines, no matter how complex they may be inside, as black boxes; that is, as devices which provide simple output given the user's input. Users rapidly learn the universe of input, reactions, and behaviors of these machines, and can easily translate that learning to new and different models and brands of what is essentially the same product, even as the interior mechanisms of these models and brands are changing substantially, rapidly, and wholly beyond our ken.

There *were* some computers which had such bounded complexity; for example, the Wang and Lanier dedicated word processors. Today's dedicated game consoles made by Nintendo, Sega, and their competitors, too, have such bounded complexity (although their new peripherals and links to the Internet threaten that stability). It is, of course, the word "dedicated" that gives away the bounded nature of these machines. The Lanier and Wang machines acted as black boxes, exhibited predictable behavior, and were easily mastered. But the dedicated systems did not last long. As single function computers, they were a reified oxymoron, and given a choice between the dedicated Wang and the IBM PC, an unrestrained computer, we chose the latter and its descendants. In the early 1980s, the IBM PC became a game machine, a communications device, a calculator, a word processor, a database manager, and importantly, the PC ran a spreadsheet program allowing not just calculation but also strategic economic planning and related statistical analyses.

Such a multifunction box could be thought of as a series of black boxes, simply many items exhibiting bounded complexity. And that may have been true for a while. But the situation changed as programs made by different companies began interacting with each other. Programs swapped data, relied for their input on other programs' output, shared subprograms known as modules or APIs (application program interface), and interacted closely with programmed and programmable hardware. Such interactions, begun within a single box, amplified exponentially in the networked environment. Users modified their store-bought programs, and consciously—and sometimes unintentionally—modified the way their programs operated. At the same time, mainstream programming languages moved to the use of object-oriented programming (OOP). These new programming languages expressly validated the reuse of the same chunks of code in a variety of computer environments and platforms for a variety of operational purposes.

These custom elements together produce unique computing environments, none quite the same as another, in which complexity is unbounded. Sitting down at a strange computer is not like sliding into the seat of a rental car or picking up a strange telephone. What happens deep inside the machine—the way its multifarious elements interact and the way it interacts with other networked machines—matters a great deal. There is no static black box here, at least not necessarily. The physical box is not even congruent with the computing environment, because that environment is dispersed among many computers across a network. Those who are more knowledgeable about the capabilities of the computer use it differently from those who are relatively ignorant. Virtually all users of consumer devices can accomplish practically the same tasks, with the same level of ease and using generally the same features. But computer users vary widely on the task they can achieve, their relative ease of use, and software features that they are able to efficiently harness.

A substantial number of people in the United States and other Western industrialized countries use computers, but they have enormously different experiences of those machines. People who otherwise think of themselves as members of the same community—from the same neighborhood, or part of a state, or workplace, or industry—will not necessarily share the same assumptions about computers or have equivalent knowledge bases. That reveals the lack of a *common* understanding about nature of computers, their workings, capabilities, and risks. These things, these computers, are different for different people.

And that is the fourth reason that the law has difficulty in finding and applying persistent analogies and metaphors from past cases to issues involving new technology.

As a result of these factors, which are explored in further detail in the essays that follow, the law is inhibited in its ability to regulate advanced computing technology, and concomitantly the legal system risks losing its moral authority to do so. In this way, the legal system is changed, as it subjects technology to its strictures.

PUBLIC AND PRIVATE LAW

Classic litigation is not the only way to resolve disputes, although I have assumed that so far. There are alternatives.

Classic litigation starts with the filing of a complaint, a series of formal allegations, in court. Most of the states and the federal system decided a few decades ago that trials should not be conducted "by ambush" or "by stealth." And so the courts provided broad rights of "discovery," a process by which one side discovers from the other side everything that might be relevant to the preparation for trial. This process usually devolves to interminable written questions, stunningly long requests for documents, and lengthy depositions, or oral question sessions, with potential witnesses and others who might have useful information. That process can take years, because often discovery battles erupt concerning the propriety of the questions and the sufficiency of the responses. This pretrial diet is spiced with motions for more—and for less—discovery, for better and clearer complaints, and sometimes motions designed to test the law's applicability to the facts.

A small percentage of these cases will go to trial. The percentage is small for reasons alluded to earlier: it is a long and expensive wait until a case can be sent out to trial, and the legal uncertainties can make a trial look like no more than dice roll. And so parties settle, maintaining some felt control over their fates.

There are, as I suggested, alternatives. Arbitration is the process by which the parties give up their rights to the public trial system and agree to be bound by the decision of a neutral arbitrator. The rulings are not appealable. The parties by contract can, in detail, set out the rules: Perhaps there shall be no discovery; perhaps the case shall be done in 150 days; maybe the arbitrator shall be someone with a minimum of 10 years expertise in the substantive legal field; the parties may agree to have hearings in Tulsa, Geneva, or London; perhaps there shall be a three-arbitrator panel—and so on.

Parties can also agree to mediation. Mediation is a settlement conference, at which a presumably trained mediator works to have the parties hammer out an agreement that is not binding until the parties say so by signing a new contract.

In California and other states, parties can "rent" a judge from a stable of retired judges and have an essentially private trial under guidelines established by the parties.

The role of these so-called alternative dispute mechanisms is to take the legal dispute away from the defaults of the public justice system and to have it ruled by private law. This is a shift of procedure, that is, private rules of procedure replace those declared by courts and the legislatures.

But the public/private dichotomy can operate substantively, as well. Those who do not like the way the public law of copyright or trademark assigns rights can create a private contract to do it differently. Employers and employees can adjust their entitlements in writing. Joint authors who do not appreciate the copyright law's per capita division of ownership rights can make their up their own percentages.

More generally, those troubled by legal uncertainty can resolve doubts in writing. They make private law because it is difficult to extrapolate the generalities of present public law to the specifics of the future. For example, doubts on whether certain communications are entitled to the protections of privacy rights can be resolved by express consensus. This private law is made on the spot, as it were; it is pertinent, fresh—certainly compared to the dusty tomes of case law.

But this resort to private, substantive law is not a panacea. Scriveners of these private contracts will do their best to forecast the development of law and technology, but they are not, obviously, always right. More importantly and subtly, the public law is still used to interpret private contracts: Public courts may become the forum for the contract fight. Words that no one thought to define (or the text of those definitions themselves) are construed as are similar terms from public statutes and cases. Arbitrators often see themselves as a variety of judge—and use the same public cases to which judges would refer.

Finally, and in sum, judges and arbitrators will generally use the same fundamental rule when faced with a dispute on the meaning of a key term in one of these private contracts: they will ask if the parties manifested, in the writing or in their actions, "objective" evidence of mutual assent. They will ask: Did the parties do something that, to a "reasonable" outside observer, manifested an agreement?

Those are the classic legal questions; but by their very terms they presuppose a communal consensus—the lack of which the private agreements were meant to offset.

This should not be utterly surprising. Even private contracts, at heart, must always be progeny of the public law and public enforcement mechanisms. Contracts are expressed in language, which is of necessity a public artifact, such that the meaning of words is generally in the public domain. And after all, what use is a contract if no court will enforce it? What good are arbitrators' judgments if the courts will not enforce those? Thus, from the perspectives of both interpretation and enforcement, public usage of terms—such as is found in statutes and case law—will play a critical role in private lawmaking.

And we should not forget the criminal law. As one of the essays here notes, state and federal criminal law have come to the electroverse with a vengeance. And this area is reserved, wholly, to the public realm: No private contract (save one for immunity, made with the prosecutor) will exempt one from the eye of that public law.

Public law cannot be displaced. The discords and treaties between the law and advanced computer technology, discussed in these essays, will not be avoided.

References

[1] de Montaigne, Michel, *Essays*, Middlesex: Penguin, 1958.
[2] "Survey of the World Economy," *The Economist* 15 (Sept. 8, 1996).
[3] "Survey of the World Economy," *The Economist* 8 (Sept. 8, 1996).
[4] "Survey of the World Economy," *The Economist* 10 (Sept. 8, 1996).
[5] "Software explosion rattles car makers," *Electronic Engineering Times* 1 (Oct. 28, 1996).

Part II
Legal Assumptions

This collection begins with a chapter on a basic legal assumption: the legal fiction of the "reasonable man." There is some irony that the law depends on a fiction, but it should be recalled that the "reasonable man" is frankly an idealized standard, no less worthy of our respect than such standards in other fields. Just as disciplines such as sports, the arts, and many sciences may be understood by their ideals, so will a look at this old doctrine introduce the law in a general way. I have, of course, chosen a doctrine that is both central to and most placed at risk by rapidly developing technology.

Legal assumptions work at many levels, and practicing lawyers do not always perceive them clearly. Unquestioned in most cases are a wide range of assumptions that will have as much to do with the outcome of a case as the express rulings of judges or the patent mandates of statutes. For example, we assume judges and juries know things that need no explanation, such as how to use a telephone, or that small children do not know how to swim, or that a lawyer who wants to ask one person questions for ten days is being unreasonable. We assume that the system has common sense, although some of these "facts" will be common only to some portion of the population. It is common sense that computer files can be altered and that the sun comes up in the east; but how common is it to know that a magnetic field can destroy data on a floppy disk? Or that email can be easily intercepted? Or that cellular phones are not secure?

Aside from factual assumptions that we do not bother to strictly prove in a case, the law has general doctrinal assumptions, too, that weigh the balance against one party or the other. These assumptions are often decisive in a case. In the United States there is a presumption against "prior restraint," such that courts are unlikely to stop a newspaper or magazine from publishing, even given serious allegations of libel or invasion of privacy. In the area of economic litigation and antitrust law, courts assume rational market forces and that companies will not usually knowingly reduce prices below cost. In the consumer law context, courts assume that contracts are voluntarily entered into, even when one party appears to have had little real economic choice. Judges assume that patents are valid. More mundanely, the law assumes that properly posted mail was received at the marked address and that a business record is usually genuine. Courts also make official presumptions that are plainly fictions: they assume that everyone has "notice" of (i.e., knows about) record title ownership of real property and that the terms of written contracts were read and understood by all parties.

For each of these assumptions, there is a corresponding burden on people who must overcome the assumption with powerful proof. And often, that burden cannot be shouldered, and the case is lost. The judicial system maintains these assumptions because it is simply impossible to reinvent the wheel, as it were, in every case. The most simple car accident trial would stretch well into the millennium if lawyers had to prove the direction the sun rose, that cars move forward when accelerators are hit, the operations of brakes and street lights, that the parties in the case were not blind, and so on. Just as we use the shortcuts of assumptions and common sense in our daily lives, so too, the courts have their assemblage of legal assumptions that allow the legal system to cut to the chase.

Problems arise for a variety of reasons. First, one can never predict with certainty that a given judge or jury will have the same brand of common sense as one's client. The meaning of words common in one industry have other meanings in other contexts, and the meanings of words change over time. Assumptions are made about specific facts and words. Lawyers may go to great pains to define the terms in their contracts in unmistakable words, but those explanatory words in turn may mutate. One of my clients assured me that *everyone* knew what a "stand alone" software product was and no further definition was needed: I was reminded of the assumption made by Intel and AMD that *everyone* knew what "microcode" meant. That assumption landed those companies in years of litigation and arbitration. What is blindingly obvious to one is obscure to another.

Second, this legal mix of factual and doctrinal assumptions is not easily sorted out and may blend imperceptibly into bias and prejudice. And sometimes this occurs not so imperceptibly: In the 1800s, courts in California and elsewhere barred Chinese immigrants from serving on juries and from testifying, assuming that they did not have the same sense of justice and truth-telling capabilities as the white settlers. Some have suggested that current assumptions about the rational behavior of corporations (e.g., in the antitrust area, that companies will not knowingly lower prices below cost without assurance that the losses will be made good when all competition is eliminated) are themselves just complex economic prejudices.

Third, we should recall that many of these doctrinal "assumptions" are the accretions of judicial opinion on judicial opinion, layered on over the years by scores of judges in scores of cases. The process is reminiscent of the way, perhaps, that memory is layered upon memories, informed by the recollections of others, recreating and rewriting history as it goes. Judges refer to each others' work, and the work of their predecessors, in the ever-evolving development of the law, but it is sheer fantasy to see this as the making of a single fabric.[1] There is a lot of wiggle room in the use of precedent, because there is so much of it, and none is ever precisely on point. There is an enormous scope for a single judge's discretion, and so judges purporting to apply the "same" law may well reach different results.

Too, precedent can perpetrate exceedingly odd rules, procedures, and presumptions that will not change until the level of frustration with the old rules reaches a breaking point.[2] As a consequence, and until that breaking point is reached, many cases will suffer under inappropriate assumptions, which have found their only life and warranty in timeless repetition.

And finally, as a general consequence, we know that some of the legal system's assumptions are not congruent with the public world the law desires to regulate. It is that discontinuity that is outlined in broad terms in the following chapter, and taken up in detail in the remainder of this volume.

References

[1] Judge Wysanski in *U.S. v. United Shoe Machinery Corp.,* 110 F. Supp. 295 (D. Mass 1953).

[2] Karnow, Curtis, "Archeology of Error: Tracing California's Summary Judgment Rule," 24 *Pacific Law Journal* 1845 (July 1993).

1. "It is delusive to treat opinions written by different judges at different times as pieces of a jig-saw puzzle which can be, by effort, fitted correctly into a single pattern" [1].

2. I have discussed a bizarre procedural rule, the result of an old error and subsequent blind adherence, in [2].

The Reason of the Law: Community Meaning in the Conduct of the Legal System

1

The legal system and the community at large enjoy a symbiotic relationship: Each sustains the other. The legal system is shot through with standards that, by their terms, are based on community-wide consensus. Legal judgments are abstracted from the plurality of persons governed by the system. The law, in turn, can be a civilizing force to the extent that it accurately extrapolates the best from the multiplicity of individuals, groups, and organizations it governs. The law both depends on an established community and assumes that it has the authority to coerce and extrapolate it.

The legal system accomplishes its role by its formal reliance on the past: tradition and precedent. In the West, the law no longer judges by default, having lost the authority of the divine, or of the king. Rather the law governs by its uses of the past, specifically acting on two of its aspects.

First, as a matter of course, we all base our actions on the past. The law as it governs future action pays strict attention to this source of habit, common sense, and induction. The law relies expressly on analogous precedent and on *stare decisis* (i.e., adherence to decided law). Second, we view the past as if through telescope: our review of necessity condenses, or precipitates and abstracts, the enduring if occasional wisdom of the past.

Powerful and credible legal judgments appear to be based on that reliance on the past, the collected knowledge and sense of the community. It is a community that, after all, is partly made of those alive but, in far greater numbers, by the dead. The law is eviscerated to the extent that those bases are replaced by transient politics, corruption, or systemic delay.

The law is a powerful weapon with which to forge consensus and ideals. It has stood for what is best and noble in our civilization, lying as it does at the heart of what it means to be civilized. And so we are deeply wounded when it fails.

And it *is* failing. Ephemeral politics and a sluggard system have taken their toll. There is increasingly less assurance that disputes will be decided on the basis of community-endorsed standards—in short, the law. Instead, lawsuits rise and fall on the tides of crowded courtrooms, swamped in legal maneuvers and minutiae often invisible to the public and ignored, in large part, by judges and juries.

Strengthening the legal system through the elimination of delays and abuses can reimpose community values on the conduct of litigation, in turn reinvigorating the community itself. The following section discusses the essential role of community-based standards in the legal system. Next, the evisceration of a cardinal standard is discussed, followed by an outline of the effect of that loss in the day-to-day conduct of litigation. The final section concludes by again linking the fate of the legal system to the strength of generalized community standards.

THE ROLE OF COMMUNITY-BASED STANDARDS

Obscenity is the best known substantive legal theme that presumes community. While most speech is forcefully protected by the First Amendment, obscenity is not. Juries in obscenity prosecutions are expressly instructed to apply local community standards in judging what is, and what is not, obscene. Thus, the test anticipates that the same pictures and words may subject the publisher to a prison term in one district and an arts medal in another.

This is the best known, and perhaps most exaggerated, example of congruence between community standards and the law. It is exaggerated because the "communities" in this context are ill defined, and small; and because obscenity is a narrow and vague issue, one on which people who generally hold similar broad values may differ sharply in specific cases of supposed obscenity.

But there are more venerable, general, and thus more powerful legal tests linked to community standards. The classic test for a *tort* (the word used to describe a private wrong) depends for its meaning on a what a "reasonable" person would do. Such a reasonable person is contemplated by the basic tort of negligence, such as negligence in the operation of a machine, or the negligent release of a dangerous animal or contaminated water onto neighboring land. In all of these cases, the judge or jury will test the actions of the defendant according to whether he acted "reasonably" (i.e., as a "reasonable" person ought to have done under the circumstances).

The "reasonable" person is a legal fiction, but theoretically a fiction only in the sense of being a composite, an agglutination of common traits and habits; this person is an embodiment of common sense and the unspoken rules we all live by. Theoretically, as I say, the "reasonable" person may be a fiction, but he is a *true* fiction, an accurate summary of what is common.

Contract disputes, too (which, with torts, comprise the usual fare of legal complaints among private parties), are often subjected to the reasonable person test. In many jurisdictions, the meaning of written contracts is subject to an "objective" test: What would a "reasonable" person think the terms meant? The court asks: Is a party arguing for an interpretation of which the written words are *reasonably* susceptible? This disposes of foolish and frivolous arguments—arguments that the community believes to be preposterous.

The legally "frivolous" is defined by contrast with the expectation of the reasonable person—or, at least, reasonable lawyer (if that is not an oxymoron). A recent United States Supreme Court ruling confirmed that a frivolous lawsuit is one which no reasonable lawyer would believe has merit. It is not enough under that test to show that one used the legal system to abuse the opponent subjectively, maliciously, and without regard to the facts. Such a lawsuit is not frivolous unless, in addition, the lawsuit fails the asserted *objective* standard of the "reasonable person."

The reasonable person makes an appearance in the criminal and Constitutional law as well, governing the relationship between the government and the citizen. We have rights under the Fourth Amendment against "unreasonable" searches and seizures, and Fourth Amendment lore defines the zone protected by the courts as that circumscribed by a "reasonable" expectation of privacy. First Amendment speech rights may be limited by regulations specifying a reasonable time and place for expression, such as in the permitting of public parades and demonstrations. Self-defense in a murder case depends on whether the accused "reasonably" thought his life was in jeopardy—whether a reasonable person under those circumstances would have pulled the trigger.

The legal fiction of the reasonable person, and of his reasonable judgments, runs throughout the law. By definition, this grounds the law on reason, making it, in a word, *just*, which is how the venerable *Black's Law Dictionary* defines "reasonable." This standard is ubiquitous precisely because it is conceived as a standard: a single and exclusive rule that expressly disclaims all subjectivity. The standard is directly juxtaposed with the motives and desires of individual actors on the legal stage and is designed to overrule personal opinion. The standard is meant to supersede an individual judge's personal foibles, just as it overrules an individual lawbreaker's individual predilections.

The standard is part of the magic and drama of the law, part of its formality. Together with black robes, legal forms, and the secrecy of judicial deliberations, the reasonable person standard helps to sanction the legal system. Its assumption of objectivity authorizes it to decide legal disputes in which half the litigants must inevitably lose but nevertheless must accept the result.

Like other objective precepts, the reasonable person standard is meant to allow prediction. A common, unitary test permits people to know in advance that their actions will, or will not, lead to liability. Without the predictive capability of legal rules, there is no sense to the notions of lawful behavior or of a lawful society.

To be sure, one cannot perfectly predict one's way through the infinite variety of factual scenarios, the endless interplay of statute, regulation, and court-made law. But we are meant to be guided through those labyrinths by general rules or principles, abstracted from the details of *sui generis* cases. The reasonable-person standard is such a principle. Based on it, we predict whether millions of dollars will be won or lost, whether a home or car can be searched by the police—sometimes, whether one is sent to the gas chamber.

THE EVISCERATION OF A CARDINAL STANDARD

The reasonable person standard is associated with the development of the tort of negligence, with negligence defined as an act or omission that a reasonably prudent person would not do. That development dates to the early 19th century. The reasonable man (and then, it *was* a man) "was the personification of the community ideal" as Professor William Prosser reminds us in his classic *Law of Torts.* This mythical man was a good citizen; reasonably intelligent; careful and attentive to his wife, his property, servants, and animals; and sufficiently schooled to know the basic laws of physics, the phenomena of nature, the way machines work, the limits of his strength, the habits of others and their propensity to crime and negligence, and the dangers of flammable liquids, electricity, and so on.

This "wond'rous" man grew up when the lawyers and judges responsible for wielding the doctrine, one supposes, had a common background from which to draw. But the American legal system, like the society in which it operates, has changed. The system appears increasingly incapable of generating a consensus on what is reasonable and what is frivolous. The most extraordinary propositions are taken seriously by judges, insupportable claims are routinely made, and lawyers are increasingly incapable of accurately predicting the outcome of cases.

The best-intentioned and capable lawyers and judges are frustrated in their efforts. They find themselves applying legal tests, such as the reasonable-person standard, that produce divers and ambiguous results. The amount of police force reasonably required to subdue Rodney King divided the country, and different emphases at trial produced opposed jury verdicts in the state and federal proceedings. In a variety of recent cases, judges and juries have differed as to whether a defendant was reasonable in the use of deadly force. And any antitrust lawyer will report profound difficulty in predicting whether a jury (or appellate judges, for that matter) will find an aggressive competitor to have "unreasonably," and so illegally, restrained trade.

The sense of an empty standard pervades the daily grist of litigation. So-called motion practice and discovery proceedings constitute the bulk of a litigator's professional life, for most cases (90% and up) are disposed of before trial. The cost of litigation is perforce due to such mundane pretrial and discovery practice. In efforts to reduce that cost and eliminate abuses, courts penalize frivolous motions and impose money sanctions. But that, in turn, has led to expensive and complex litigation on the propriety of such penalties.

And so a new area of legal practice has been born. Special magazines and case reporting services, self-professed specialists, and a mountain of appellate authority have been unleashed, all speaking to the imposition of sanctions for frivolous tactics. In many cases, there is no motion which does not also include a side dispute on sanctions for asserted frivolous tactics. Those disputes, impugning the integrity and pocketbooks of the lawyers themselves, are often fought with more zeal that the underlying legal battle.

Lawyers argue about this, at great length and expense, because judges allow them to argue about it. There is little, if any, consensus on what is frivolous.

THE LAW UNCHECKED BY A COMMON SENSE

As an icon of the law, the reasonable-person standard has a binding force. It suggests fairness, reason, and a predictive authority by which we may organize our lives. But, like all symbols, the standard is influenced by the realities it seeks to govern. The legal fiction of the reasonable person is true to the extent he really does embody the community's expectations. Where those standards are in disarray, or where the community is sundered from the legal system, then the symbol is hollow.

Which is it? Is the compact of shared meaning disintegrating among those of the community at large? Or are judges and lawyers too far removed from the citizenry? The options are not mutually exclusive. Bonds among the members of a community surely are a function of the extent to which subsets of the community sever themselves out. Separatist movements of any shape and brand, conscious or unconscious, weaken us all. Such separatism is evident not only in extreme racial and cultural contexts, where it is conscious, but also in the ordinary political landscape. It is unconsciously furthered by one-issue groups (such as pro- and antigun lobbies, prochoice and prolife groups, labor, and insurance), a byproduct quite independent of the specific positions espoused by those groups.

Professional groups, too, can mutate into separatist movements. They do so through specialized education, jargon, history, clothing, professional publications, and trade organizations. The legal profession, specifically, risks the separatist label precisely because the law depends for its authority on the image of objective, thus distanced, justice. It may be a matter of degree, but the distance between legal practitioners and the community at large is now too much. And it is more that distance, and less any dissipations of the culture generally, which produces the twin symptoms, a failure of consensus on what is frivolous, and an inability to predict the outcome of litigation.

There are many indications of the law's separation from common sense. The pervasive (if involuntary) funding of litigation by insurance companies and the eternal search for the "deep-pocket" defendant, warp the decisions of whom to sue and how to conduct a defense. The infections of politics, in its worst sense, place on the bench those whose qualifications include the payment of money to candidates and political parties.

The courts are more crowded: their number has not kept up with population increases and the criminal docket has soared. On the civil side, procedures have developed, cancerlike, which authorize endless pretrial investigations into others' testimony and documents. Those procedures are essentially ungoverned: Judges have no time, and generally less interest, in overseeing pretrial disputes. The cost of litigation typically keeps the courts for the wealthier segments of the society and precipitates settlements that reflect the cost of litigation, as opposed to the merits of the case.

Civil litigation thus is more of a private fight and less a presentation to the public's jury, checked by the public's judge. The process is usually incoherent to the public, including clients. Few cases end up before a jury, and judges rarely interfere along the way. The end result of the process is more a function of the rules and costs of the process than of the real-world controversies that precipitated the lawsuit. In a phrase, there is no *reality check.*

In such a context, an appeal to reason—that is, to the reasonable-person standard or, more simply, to a reasonable argument—can be ineffectual. The "reasonable person" is superfluous, his embodiments withdrawn. Eventually, from disuse, he vaporizes.

It is important to note that these infirmities are outgrowths of what I have termed the magic and drama of the law. Formalities, pretrial discovery and motions, insurance coverage, the imposition of money sanctions, and the ability of any citizen to sue General Motors, Nintendo, or the United States government for a million dollars—all of these have the best of intentions. These characteristics of our legal system make it equally accessible to all. But all values evolve, and like a caboose cut away from the diesel, legal values have evolved unchecked.

TOWARD COMMUNITY STANDARDS

No revolution is needed to harness the legal system back to the community it is designed to serve. Our guardian, the reasonable person, can be reinvigorated by subjecting the actions of counsel to the scrutiny of the public, for there is no other source of the reasonable-person standard. Here, the public acts through its learned surrogates, the judge and jury.

First, we cannot afford to have judges selected for their political muscle or personal contacts; the public is served not at all when judgeships are plums awarded according to hidden criteria. The courts must be staffed with men and women of broad learning and experience. They must be more than bland legal technicians; they must be utterly different from the tired and cynical judicial officers who believe they are just another player in a stylized drama, divorced from common sense and common conviction.

More centrally, every case should be filed with the expectation that a jury will probably pass on it—and soon. This is just the opposite of the current situation, where counsel now can assume that years will pass before the case is resolved and that they will never face a jury. There are places where this velocity is achieved: a district court in Virginia is known as the "rocket docket" with about six months between instigation of suit and a trial date. In such a brief period, the opportunity for abuse is much reduced: The immediacy of a trial and of jury review concentrates the mind of counsel wonderfully on the true worth of a case.

Short-fuse trials generally cannot be effected under present circumstances, because there are not enough judges or courtrooms. There are few special interests lobbying the legislatures to increase funding for the courts. This third coequal branch of government secures but a small fraction of the government's budget (about 1.3% of the state budget in California), and court funding lags behind population growth. The penalty for underfunding, though, is felt by the whole public, because a dysfunctional judicial system eviscerates all business interests, civil rights, and efficiency in the imposition of the criminal code.

To be sure, the legal system needs other palliatives as well. Legal education is administered by professors with little practical experience; those who do teach the *practice* of law are treated as pariahs at law schools. Bar associations do little to police their own members. Alternatives to the gladiatorial litigation model—such as mediation and arbitration—are poorly understood by judges and lawyers, not to speak of clients. And so on. A panacea is not offered here, only a few remedies addressing the law's general claim to authority and reason.

Of course, not every case can be ready for trial in six months. Not every infection of the judicial system can be cured by newly robust community standards. And it appears tautological to suggest that community standards themselves can be rewoven just by subjecting the legal system to those standards.

Nevertheless, speedier justice and judgments in harmony with such community values as we can muster must burnish, at least a little, the image of justice and its embodiment—the standard of the reasonable person. And the strengthening of that symbol would, in turn, vitalize the community at large.

Part III
The Collision of Law and Technology

Part III collects essays on the mutual impact of law and computer technology. Both law and technology manifest symptoms of the collision. The law remains the focus here, though, and it is that specific area of the law known as *intellectual property* that is generally most at stake.

The reader may have little substantive knowledge of the key areas of law that usually pertain to advanced computing technology. Those areas reside within the perimeter of intellectual property law and consist of trade secrets, patent, copyright, and trademark. There are related areas, too, such as *unfair competition* and breach of contract (such as license agreements that cover the use of intellectual properties). All these areas may pertain to software and to software entombed in silicon (otherwise known as "hardwired" code—or hardware) and locked into chips that process data. The following is a brief summary of a few of these areas.[1]

TRADE SECRETS

Certain aspects of software can be protected through use of trade secrects and contractual licensing agreements. The law typically requires the party asserting a trade secret right to take reasonable steps to prevent the public disclosure of the information held as a trade secret. Accidental or other public disclosure of a trade secret may eliminate the protection. Without such disclosure, the trade secret rights will remain effective indefinitely, and anyone stealing the secret can be sued and stopped.

Trade secret rights can be enforced against parties that unlawfully obtain the information held as a trade secret. It is not always simple to

1. Some of the material here on trade secrets, copyright, and patents comes from an official source: The U.S. Copyright Office.

know what "unlawfully" means in this context: unlawful misappropriation means obtaining something one knew was trade secret, from a place that one knew was meant to be concealed. Picking up a "secret" formula from the sidewalk is not illegal misappropriation. Breaking into a locked safe is.

Almost anything can be a trade secret, as long as it secures a competitive advantage for its owner by being kept secret. Secret formulas are obvious examples, as are customer lists and source code. But so is a secret seating arrangement, or color painted on the wall, if the arrangement or color (for example) boosts productivity and no one outside the company knows about it. Even the use of elements in the public domain can constitute a trade secret, if the assembly and use of the elements are secret.

PATENTS

Patents can be used to protect processes implemented using software, as well as computer-based systems. Patents are available for "any new and useful process, machine, manufacture, or composition of matter, or any new and useful improvement thereof." The invention must be new and not obvious to a person familiar with the technical field of the invention. Patent protection allows the patent holder to preclude others from making, using, or selling the patented invention for a statutory period. The party granted a patent must take action to enforce rights provided under the patent—the issuance of a patent does not automatically preclude infringing activity. Unlike copyright protection, which just makes copying illegal, patents bar *anyone* in the United States from practicing the invention even if they came up with it on their own. Patents have been issued for a wide range of computer-related inventions, such as a three-dimensional cursor, a means of encrypting data, a home video game joystick, and techniques of delivering multimedia. For more on software patents, see [1].

TRADEMARKS

Trademarks identify the origin of the goods, and can be used by consumers as indicia of authenticity and corresponding quality of goods. Trademarks can include words, phrases, numbers, unique designs, colors, or sounds. Examples of text marks are "Exxon" and "Disney." Phrases include "Fly the Friendly Skies," and the "You're in Good Hands" slogan used by a certain insurance company. We are all familiar with unique designs or logos used by many companies, such as the NBC peacock, the

AT&T logo, or the McDonald's golden arches. Harley Davidson motorcycles has applied for a trademark on the reputedly distinctive sound of the Harley Davidson engine, and distinctive color hues of pink and blue have been held to be trademarks for a certain kind of insulation and plastic film products, respectively.

Trademark protection may also be available for the "dress" or appearance and configuration of a product or its packaging. For example, the distinctive shape of the Coca-Cola bottle or certain food containers or picture frames qualify for such protection. Certain fictive figures, too, may act as trademarks: Sonic the Hedgehog is used by Sega as a trademark, as are many of the Stars Wars characters for LucasFilm. Unlicensed use of any of these slogans, colors, sounds, figures, characters, or logos violates federal trademark law (and often state law as well).

COPYRIGHT

Software code, and most other creations such as texts, graphics, and compositions are protected under copyright law as an original work of authorship. Copyright protection stems automatically from the act of fixation of a work onto a tangible medium. A copyright gives its owner the ability to control the reproduction, adaptation, public distribution, public display, and public performance of the software code. Copyrights can be used to prevent others from copying the work. Copyright owners can also prevent the unauthorized rental of software. Copyright protection cannot be used to prevent the use by others of the functional aspects of software, nor (unlike patent protection) can it be used as a basis for action against independently developed software.

Copyright law bars the unauthorized reproduction of protected works, including the public performance or display of a work and including the making of a *derivative* work from the original. A derivative work can cut across traditional disciplines: a ballet can be derived from a film, a film can be derived from a book, and vice-versa. Clothing may "derive" from home video games and/or a graphic creation, and so on. Those who own the copyright to the original own the right to bar others from making derivative works. Copyright is usually owned by the creator or author of a work, unless the author is an employee and creates the work in the course and scope of his employment, in which case the work is owned by the employer. Most anything that can be embodied or that is perceived by the senses (even if a machine such a computer is needed to assist the perception) can be the subject of copyright.

♦ ♦ ♦

> *There is no life today without software.... The world would probably just collapse.*
>
> —*James Gleick [2]*

There are a host of other legal areas that are implicated by computer technologies, and these are briefly described next in context. Privacy, free speech interests guaranteed by the First Amendment, and the criminal law may be implicated, as well rights to publicity, contract rights, the ability to sue others for their negligence, recklessness, intentional infliction of emotional distress, and indeed virtually any right recognized in other contexts. As we shall see, the computer has opened a virtual world, and the working assumption is that actions in the "real" world will have their analog in the digital realm, consequently leading to the import of the rules governing the "real" world. That is one conceit, and it has some explanatory power. It is also true that the explosion of computers, and the concomitant importance of the legal issues associated with computers, is the consequence of their use as *communications* devices, real-world communications among people in the real world. Crimes, torts, breaches of contracts, and violation of civil rights all require a means of communication but can all be accomplished without regard to the specific means of communication. Increasingly, our speech and actions are computer mediated, and so increasingly the computer finds itself at the center of myriad legal issues, not just the classic intellectual property issues alluded to earlier.

The essays in this and other sections explore some of those issues, with some attention to the question of why the computer, as opposed to telephones, paper and pen, typewriters, acoustic speakers, and other technologies of composition, authorship, and communications, should have such a powerful effect on attempts to utilize general and traditional law.

References

[1] Linck, Nancy J., and Karen Buchanan, "Patent Prosecution for Computer-Related Inventions: The Past, the Present, and the Future," 18 *Hastings Comm/Ent L.J.* 659 (Summer 1996).

[2] Gleick, James, "Little Bug, Big Bang," *The New York Times Magazine* 38, 40 (Dec. 1, 1996).

Molten Media and the Infiltration of the Law

2

The legal system infiltrates technology, like a thin mist seeping under the door, staining it and turning technology into a different animal. Programmers wonder if the code they write was patented by someone else. Graphics lose their innocence and look like trademarks, trade names, and logos; they have that old "look and feel" of someone else's product. Trademark law reaches out its sticky hand to embrace color, sound, the overall appearance of every product and packaging on the market. Copyright law hovers like a specter, infecting every line of code, every data structure, every animation, every sound, graphic, and screen layout.

And while this law spreads, it thins out, too. The types of properties that the law protects now were unknown just a few decades ago. There is an unreality to the transient audio-visual image, an insubstantiality to a user interface. There is something very peculiar about patenting a three-dimensional cursor or a software-retrieval system. Surely these are created things, to be protected from theft; but most judges are reluctant to tread too heavily here. These judges remember "property" as tangible land, gold, or cattle; this new stuff looks ephemeral. Judges have a hard time giving a user interface the same absolute, exclusive protection provided to a house, a car, or money.

So the law expands its reach to govern the development of advanced technologies, but at the same time its touch grows more hesitant and uncertain. It is like an omnipotent Imperium in nominal control of vast territories—every movement of the inhabitants may be the subject of Imperial decree, but none knows which actions, precisely, will invoke the Imperial attention.

This chapter discusses the apparent paradox of comprehensive but uncertain protection for intellectual property and the consequences for companies engaged in high-tech development.

INTELLECTUAL PROPERTY: A SPECTRUM OF PROTECTION

Imagine two kinds of creation. One covers highly creative works: the bizarre, the remarkable, the arbitrary, the unexpected, the flights of fancy and fancy footwork, the stuff that can only spring unbidden from the human mind. And then the opposite: facts, pieces of the natural world, the stuff we find all around us—grains of sand, oxygen, rocks, the patrimony of us all.

Now, the law protects the new and artificial by granting the creator a monopoly: no one else can use or have the creation. By contrast, the public and natural domains are for everyone. Copy a painting or reproduce a book, and go to jail (or worse: get sued). But take a picture of a rock or leaf, or make an imitation of the night sky, and you will be left in peace. Invent the integrated circuit chip or the process by which rubber is vulcanized and get a patent; but if one comes up with a new algorithm or discovers a physical law, no protection is available for the discovery—and everyone can use it.

In truth, there are not two types as much as two ends of a spectrum. Some work is plainly and dully derivative or duplicative of physical reality; this secures little protection from the law. Other work is stunningly unexpected, redefining a field or creating an industry, and that is strongly protected.

Trademark law participates in this spectrum of creativity. Powerfully novel terms are strongly protected, and weaker ones, less so. So, for example, arbitrary words like *Exxon* and *Kodak* are given great deference by the courts; generic terms—those that are most descriptive or most closely and obviously allied with the physical and natural worlds—are given little weight and are difficult to enforce. Phrases such as *The Hot Dog House*, *Aviation Services*, and *Drawing Program* are useless as trademarks for the goods and services they describe. But, of course, one can be arbitrary with natural things: Just as a new collage can be made of wood and paper, so, too, a good trademark can be—*Apple*, or *Virgin Airlines*.

Patent law has analogous rules, such a the doctrine of equivalents and the "means plus function" test. Both of these can broaden or narrow the effective scope of claims in a patent. Both doctrines try to discern the difference between a truly new invention that deserves a patent and the insubstantial variation on an older invention that deserves no protection; indeed, making these "insubstantial variations" would be an illegal infringement of the old patent.

Copyright law lies on the spectrum as well. Factual compendia, like telephone listings, are very hard to protect. Depictions for which there is very little choice or creativity—such as showing a pipeline on a topographical map—secure very little or no protection. Creations whose structure is dictated by physical requirements or standards are difficult to protect. For example, computer code that issues obvious and necessary calls to peripheral systems or polls input devices is not easily protected: these works are a direct function of the physical contexts and so share in the very weak—if any—protection we give to asserted rights in the real, physical world. Anyone can use them.

Normally, anything that is "substantially similar" to the original is enough for copyright liability. "Substantial similarity" catches many rough approximations—it provides a thick shield around the original work. This then establishes powerful protection for the original creation. Anything that comes close is an infringement.

But where the original's originality is at a bare minimum and creativity at its lowest tide, the law provides nothing but "thin protection." The new replicator will be punished only he makes a *virtually identical* copy. When judges apply the virtual identical test, very small differences—perhaps just a little morph or tweak—will be enough to distinguish the new work from the original; that will insulate the new work from legal attack.

The law usually has had little trouble applying this spectrum of creativity, because it has been simple to distinguish the endpoints. At the one end we have the common, universal, natural, factual, historical, physical background (that is weakly, "thinly," protected); at the other end, we have the artificial figments of the imagination, which are powerfully protected.

MOLTEN MEDIA

But our two endpoints are melting and intersecting with each other. Nowhere is this conflation more apparent than in multimedia, and the same events will erupt shortly in the budding virtual reality industries.

Developing out of computer software, classical entertainment, and video games, multimedia has begun to subsume such other trades as advertising, film, and publishing. The markets for multimedia productions are expanding rapidly, while the cost of production—authoring tools and CD-ROM publication technology—is dropping precipitously. Development tools are increasingly easier to use.

Distribution of multimedia productions is not as widespread as TV or indeed computers generally—until recently, most PCs did not even have CD-ROM drives—but those numbers are increasing dramatically. And Internet access (which is built into the current generation of PC operating systems such as Apple's System 7, Microsoft's Windows 95, and IBM's OS/2 Warp) will further the spread of the product. This is especially true with the advent of graphical Internet access, such as those provided by Netscape Navigator and Microsoft Explorer, and other developments, such as Microsoft's Blackbird, the OLE-based tool that allows Internet access to object-oriented compound documents with text, audio, and visual components. Widespread distribution of multimedia products is accelerated by the advent of Sun's Java programming language, which allows the Internet broadcasting of small multimedia programs known as *applets*.

Widespread distribution and ease of manufacture are important only when conjoined with one more element: digital production. Digital production implies that one product is easily morphed into another; it means that anything, originally digitally created or not, can in seconds be scanned, clipped, and cropped into a new product. For the digital creator, there is no difference—at all—between *creation* and *copying*.

To be sure, we all stand on the shoulders of giants; we legitimately use ideas and raw material from our surroundings and history. But before digital production, that raw material was filtered through the imagination of the mind—which, in a sense, made it the property of the new creator—*before* it was used in the new product. Even if inspired directly by an old painting, hearing an old poem, or reading an old play, the new creator would first *abstract* the old item and apply some minimal amount of new mental energy before the new work issued forth. So the new "copy" was in fact different from the original and encompassed new creativity.

Now, the literal replication can happen *first*, without the intervention of abstraction—without new creative processes. Digital replication is, by definition, *exactly* like the original. Now, we're not just standing on the shoulders of giants—we're using their legs to walk, as it were. The new means of production create enormous copyright problems.

So, too, the new means of transmission. Passing a book around from hand to hand is innocent and violates nobody's rights. But multiple browses through a file creates a series of copies and may violate the copyright law, just as a series of passes through a networked multimedia database probably generates a series of "performances" of a work that may not be authorized.

THE REACH TO REALITY

Increasingly, the "world" that lies behind and supports our creativity is a *synthetic* world. Our social and business environments are, increasingly, *electronic* worlds. Advertising, publishing, politics, and our very day-to-day life increasingly quote the electronic worlds of television, film, and video games. The encroachment of such terms as cyberspace, virtual reality, and networks all point to the gradual displacement of the so-called natural world by the digitized fabricated creations of humans.

> Someday soon, online computing will be the sea we all swim in, and when that happens, it will be the fish—the colorful, complex organisms we are beginning to call "content"—that matter most [1].

Indeed, the only "fish" will be electric ones; but unlike real ones, someone owns them.

In this new world, "fidelity to reality" takes on a very peculiar twist. The sound of real gunshots is too tame; we need fabricated sounds that everyone who watches action films will recognize as "real." References to social reality are incomplete without citations to popular television shows and movies, every one of which is the subject of copyright and trademark law. The holy grail of increasing realism is aided and abetted by the movement of our day-to-day reality towards the electronic realm: reality is electrified, and electronic reality becomes more vivid, more tangible, more visceral, more—real. The "culture *becomes* simulation" [2].

This is the description of a conflation, a collapse of two realms. And the legal world is not immune. These developments wash out the spectrum that distinguished the protected innovation—the valued, novel, new, creative invention—from the unprotected fabric of physical reality. The fusion of the real and electronic worlds destroys the frames, the contexts, the signals by which we judge whether something is true or false, real or imaginary, a joke or serious.

For example, the interface for a virtual reality battlefield simulation and a real Abrams tank—the kind that actually kills—is about the same. The difference between news and advertisements is very thin in the well-done infomercial. A "realistic" space wargame is more likely to rely on fidelity to Star Wars® and Star Trek® than NASA's ungainly contraptions. And a "realistic" car race game through downtown New York needs to show the signs and products and buildings, every one of which is trademarked and copyrighted and—for all we know—the subject of a design patent.

The new electronic reality from which we derive our inspiration and raw materials is a minefield of proprietary rights. It's as if God copyrighted the universe.

OWNING REALITY

Who does own the chunks of our routine electronic world? And how strong is the protection? Can the terms *Windows* and *Internet* be trademarked? Can a three-dimensional pointer really be patented? Can Lotus really copyright a basic spreadsheet design and interface? Can I stop others from cutting and pasting my posting on electronic bulletin boards?

Suppose I create a three-dimensional orange with Sense8's virtual reality development tools. The orange is complete with density, shape, surface resilience, pitted surfaces, and sound files clipped from a software package (for that *schhhhlerip* sound of an orange squeezed), using textures scanned in from a *Life* magazine photograph. The orange floats there, visible, tangible, as real as real, indistinguishable in an electronic realm from a real orange; as real as rocks and air and stars.

Who owns the orange? Or do different people own different parts of it? Can I give you a copy? Can I sue someone who takes the orange code to incorporate it in another environment?

There is no part of this virtual orange that is not subject of legal claims. Separately, the code, the *structure* of the code, perhaps the data file structures, the sound and image bit-mapped files, are all copyrightable; patent law might cover some portion of the code, and certainly some portion of the authoring tools and display hardware. And who knows, perhaps some company—Orange Julius?—will bitterly complain that its trademark is being taken in vain, diluted in its worth if not with water.

The law will indeed find enforceable rights in these contexts. But we do not know which end of the spectrum of protection will predominate. The courts may find these chunks of the electronic world deserving only of thin, weak protection, which would allow a relatively high degree of freedom in the uses to which these properties are put by others. Or the courts may invoke strong protection, severely restricting use by others.

Because the environment is an artifice, the legal system claims a controlling interest. But as it does so, these legal doctrines of ownership begin to dissolve in uncertainty. At the same time, the legal system itself—the mechanism of decision—has become clumsy and intolerably expensive. And as the legal system unravels, it threatens to take the electronic world—our world—with it.

THE INFIRM LAW

No one usually gets to trial within a year, and getting there can be stunningly expensive: Routine motions can cost $15,000; many experts charge

$300 to $400 an hour; lawyers charging $290 an hour can spend days traveling, preparing for, and taking a long series of sworn statements (depositions). It adds up quickly. Some cases involving intellectual property are brought in state courts, under the guise of state law; but the state courts in the major metropolitan areas are grossly overworked, underfunded,[1] and rarely staffed with judges familiar with advanced technology (there are exceptions). In some states, it can take four years or more to get to trial. Federal courts may be faster, but they are not cheaper.

Intellectual property litigation takes a long time; it is unpredictable; and it is ferociously expensive. This is just the sort of activity that can quickly kill off the small advanced technology company. It is the land of the Pyrrhic victory: one may be innocent of infringement, but it will cost $350,000 to defend the suit. It is no wonder that less than 5% of cases get to trial. The rest, if not dismissed after extensive evidence-taking and motions, are settled. The terms of those settlements reflect less about the merits of the suit and more about the realities of the time and expenses of litigation.

Litigation here is an assault weapon. Fear of litigation deters the new creators, artificially expanding the effective monopoly power of older intellectual property.

We have then a potent combination: An infirm legal system ramifying throughout the digital reality that envelops us. We have bred an artificial bacillus that proliferates well in the synthetic reality.

CONCLUSIONS: CLIMBING OUT OF THE LEGAL PIT

The explosion of the digital context takes the concept of *property*, an integral aspect of creation and invention, and ramifies the notion of "property" throughout the mundane electronic world. This invocation of "property" inevitably brings with it the baggage of a hobbled, underfunded, and heavy-handed legal system. Every virtual gesture, each read and write-to data, each poke and peek, each shift in the electronic world teems with legal implications. Those implications are complicated and ambiguous because the legal system, as a whole, is splintered.

Fundamentally, the effect is stultifying.

Reforming the legal system would help. Funding the courts at acceptable levels and increasing the quality and quantity of judges would cut down on the interminable delays and consequent uncertainties. The legal system would do better with more judges with the courage to swiftly

---

1. For example, the judiciary in California—the third branch of government—secures about 2% of the state's budget.

and effectively punish those who abuse procedures—judges who do not think of a lawsuit as simply a sophisticated game. Clarifying and simplifying the controlling statutes would help, too, of course.

But those are long-term goals; civil justice reform is a long, slow, grinding process. In the meantime, the manipulators of high technology need a way to navigate out of the miasma of law.

To do so, one must bypass the courts and their attendant pretrial procedures. So-called alternative dispute resolution mechanisms, including private arbitration and mediation services, are increasingly available. The World Intellectual Property Organization provides arbitrators skilled in intellectual property, and agencies such as the American Arbitration Association can provide expertise as well. Too, new groups designed to provide these alternatives to specific industries, such as multimedia, are springing up. In each case, these alternative procedures require us not to reach for the heavy weapons; to leave the formal legal process alone. That is not always a temptation that can be resisted.

Underlying the resort to the formal legal system is the unspoken notion that all injury must be redressable. We are quick to perceive affront and quick to insist on the perceived right to sue. In the manmade domain of the electronic sphere, there are, surely, no accidents, no forces of nature: human fault surely must underpin every indignity. It is tempting to think so, for if no human is at fault—then what *is* the cause?

But not every injury should be the subject of suit. Accidents and other events beyond the intent of humans happen in the natural world, and no one, we hope, sues for those. As we complete our movement from citizens of the natural, to the mechanical and now the electronic worlds, so too may we begin to see the events in the electronic domain as beyond the grasping reach of the law. We may begin to see the electronic world as "owned" by all of us; we may come to see the design of new works as dictated by the public requirements and standards of virtual worlds, and therefore relatively safe from legal attack. We may come to think of objects in cyberspace as the patrimony of us all. That, more than anything else, will lose the bonds of an awkward and ailing legal system, and allow creativity the freedom it needs to copy from its environment.

References

[1] Jennings, Charles (cofounder of the Oregon Multimedia Alliance), in "Some Brief Glimpses at the Online Future" by David Batterson, dbatterson@ATTMAIL.COM (Internet download).

[2] Darley, Andy, "From Abstraction to Simulation: Notes on the History of Computer Imaging," in *Culture, Technology, and Creativity in the Late Twentieth Century,* P. Hayward (ed.), London: Libbey.

Bibliography

Articles Dealing With Cyberspace

Arnold-Moore, Timothy, "Legal Pitfalls in Cyberspace: Defamation on Computer Networks," available at http://www.kbs.citri.edu.au/law/defame.html

Beall, Robert, "Developing a Coherent Approach to the Regulation of Computer Bulletin Boards," 7 *Computer/L.J.* 499 (1987).

Beck, Henry, "Control of, and Access to, On-Line Computer Data Bases: Some First Amendment Issues in Videotex and Teletext," 5 *Hastings Comm/Ent L.J.* 1 (1982).

Becker, Loftus E., Jr., "The Liability of Computer Bulletin Boards for Defamation Posted by Others," 22 *Conn. L. Rev.* 203 (1989).

Becker, Lynn, "Electronic Publishing: First Amendment Issues in the Twenty-First Century," 13 *Fordham Urb. L.J.* 801 (1985).

Berman, Jerry, and Daniel J. Weitzner, "Abundance & User Control: Renewing the Democratic Heart of the First Amendment in the Age of Interactive Media," 104 *Yale L.J.* 1619 (1995).

Branscomb, Anne W., "Anonymity, Autonomy and Accountability: Challenges to the First Amendment in Cyberspace," 104 *Yale L.J.* 1639 (1995).

Brooks, Thomas D., "Catching Jelly Fish in the Internet: The Public-Figure Doctrine and Defamation on Computer Bulletin Boards," Note, 21 *Rutgers Computer Technology L.J.* 461 (1995).

Burk, Dan L., "Patents in Cyberspace: Territoriality and Infringement in Global Computer Networks," 68 *Tulane L. Rev.* 1 (1993).

Burk, Dan L., "Trademarks Along the Infobahn: A First look at the Emerging Law of Cybermarks," *Richmond Journal of Law & Technology,* (1995) available at www.urich.edu/~jolt/v1i1/burk.html

Byassee, William S., "Jurisdiction of Cyberspace: Applying Real World Precedent to the Virtual Community," 30 *Wake Forest L. Rev.* 197 (1995).

Cangialosi, Charles, "The Electronic Underground: Computer Piracy and Electronic Bulletin Boards," 15 *Rutgers Computer & Tech. L.J.* 265 (1989).

Cavazos, Edward A., "Computer Bulletin Board Systems and the Right of Reply: Redefining Defamation Liability for a New Technology," Note, 12 *Rev. Litig.* 231 (1992).

Charles, Robert, "Computer Bulletin Boards and Defamation: Who Should Be Liable? Under What Standards?" Note, 2 *J.L. & Tech.* 121 (1987).

Comment, "On the Service Providers & Copyright Law: The Need for Change," 1 *Syracuse J. Legis. & Policy* 197 (1995).

Conner, David J., "*Cubby v. CompuServe*, Defamation Law on the Electronic Frontier," 1 *Geo. Mason Ind. L. Rev.* 227 (1993).

Cutrera, Terri A., "The Constitution in Cyberspace: The Fundamental Rights of Computer Users," 60 *UMKC L. Rev.* 139 (1991).

Cutrera, Terri A., "Computer Networks, Libel and the First Amendment," 11 *Computer/L.J.* 555 (1992).

Di Cato, Edward M., "Operator Liability Associated with Maintaining a Computer Bulletin Board," 4 *Software L.J.* 147 (1990).

Dierks, Michael H., "Computer Network Abuse," 6 *Harv. J.L. & Tech.* 307 (1993).

Di Lello, Edward V., "Functional Equivalency and Its Application to Freedom of Speech on Computer Bulletin Boards," 26 *Colum. J.L. & Soc. Prob.*, 199 (1993).

Dunne, Robert L., "Deterring Unauthorized Access to Computers: Controlling Behavior in Cyberspace Through a Contract Law Paradigm," 35 *Jurimetrics J.* 1 (1994), available at www.cs.yale.edu/pub/dunne/jurimetrics/jurimetrics.html

Durant, Alan "A New Day for Music? Digital Technologies in Contemporary Music Making," in *Culture, Technology, and Creativity in the Late Twentieth Century*, P. Hayward (ed.), London: Libbey.

Elkin-Koren, Niva, "Copyright Law and Social Dialogue on the Information Superhighway: the Case Against Copyright Liability Against Bulletin Board Operators," 13 *Cardoza Arts & Ent. L.J.* 345 (1995).

Faucette, Jeffrey E., "The Freedom of Speech at Risk in Cyberspace: Obscenity Doctrine and a Frightened University's Censorship of Sex on the Internet," Note, 44 *Duke L.J.* 1155 (1995).

Faucher, John D., "Let the Chips Fall Where They May: Choice of Law in Computer Bulletin Board Defamation Cases," Comment, 26 *U.C. Davis L. Rev.* 1045 (1993).

Fiss, Owen, "In Search of a New Paradigm," 104 *Yale L.J.* 1613 (1995).

"Fourth Annual Benton National Moot Court Competition: System Operator Liability for Defamatory Statements Appearing on an Electronic Bulletin Board," 19 *J. Marshall L. Rev.* 1107 (1986).

Garramone, Gina M., et al., "Uses of Political Computer Bulletin Boards, 30 *J. of Broadcasting and Electronic Media* 325 (1986).

Gilbert, Jonathan, "Computer Bulletin Board Operator Liability for User Misuse," Note, 54 *Fordham L. Rev.* 439 (1985).

Goldstone, David, "The Public Forum Doctrine in the Age of Information Superhighway (Where are the Public Forums on the Information Superhighway?)," 46 *Hastings L.J.* 335 (1995).

Hardy, I. Trotter, "The Proper Legal Regime for 'Cyberspace,'" 55 *U. Pittsburgh L. Rev.* 993 (1994).

Jensen, Eric C., "An Electronic Soapbox: Computer Bulletin Boards and the First Amendment," Comment, 39 *Fed. Comm. L.J.* 217 (1987).

Johnson, David R., and Kevin A. Marks, "Mapping Electronic Data Communications Onto Existing Legal Metaphors: Should We Let Our Conscience (and Our Contracts) Be Our Guide?" 38 *Vill. L. Rev.* 487 (1993).

Kahn, John R., "Defamation Liability of Computerized Bulletin Board Operators and Problems of Proof," available at www.eff.org/pub/Legal/bhs_defamation_liability.paper (1989).

Karnow, Curtis, "Data Morphing: Ownership, Copyright, and Creation," in this volume.

Karnow, Curtis, "The Encrypted Self: Fleshing Out the Rights of Electronic Personalities," in this volume.

Katsh, M. Ethan, "Cybertime, Cyberspace and Cyberlaw," 1995 *J. Online L. Art.* 1.

Katsh, M. Ethan, "The First Amendment and Technological Change: The New Media Have a Message," 57 *Geo. Wash. L. Rev.* 1459 (1989).

Katsh, M. Ethan, "Rights, Camera, Action: Cyberspatial Settings and the First Amendment," 104 *Yale L.J.* 1681 (1995).

Krattenmaker, Thomas G., and L. A. Powe, Jr., "Converging First Amendment Principles for Converging Communications Media," 104 *Yale L.J.* 1719 (1995).

Laden, Mark L., and Stephanie E. Lucas, "Comment, Accidents on the Information Superhighway: On-Line Liability and Regulation," *Richmond Journal Of Law & Technology*, available at http://www.urich.edu/~jolt/v2i2/caden-lucas.html (1995).

Lemley, Mark, "Shrinkwraps in Cyberspace," 35 *Jurimetrics J.* 3 (1995).

Lemley, Mark A., "Rights of Attribution and Integrity in Online Communications," 1995 *J. Online L. Art.* 2.

Lessig, Lawrence, "The Path of Cyberlaw," 104 *Yale L.J.* 1743 (1995).

Long, George P., III, "Who Are You? Identity and Anonymity in Cyberspace," Comment, 55 *U. Pitt. L. Rev.* 1177 (1994).

Loundy, David J., "E-Law: Legal Issues Affecting Computer Information Systems and System Operator Liability," 3 *Alb. L.J. Sci. & Tech.* 79 (1993).

Loundy, David J., "E-Law 2.0: Computer Information Systems Law and System Operator Liability Revisited," available at www.eff.org/pub/Legal/e-law.paper (1994).

McDaniel, Jay R., "Electronic Torts and Videotex—At the Junction of Commerce and Communication," Note, 18 *Rutgers Computer and Tech. L.J.* 773 (1992).

McGraw, David K., "Sexual Harassment in Cyberspace: The Problem of the Unwelcome E-Mail," Note, 21 *Rutgers Computer Technology L.J.* 491 (1995).

"The Message Is the Medium: The First Amendment on the Information Superhighway," Note, 107 *Harv. L. Rev.* 1062 (1994).

Miller, Philip H., "New Technology, Old Problem: Determining the First Amendment Status of Electronic Information Services," Note, 61 *Fordham L.R.* 1147 (1993).

Naughton, Edward J., "Is Cyberspace a Public Forum? Computer Bulletin Boards, Free Speech, and State Action," Note, 81 *Geo. L.J.* 409 (1992).

Perritt, Henry H., Jr., "Dispute Resolution in Electronic Network Communities," 38 *Vill. L. Rev.* 349 (1993).

Perritt, Henry H., Jr., "Introduction: Symposium: The Congress, the Courts and Computer Based Communication Networks: Answering Questions About Access and Content Control," 38 *Vill. L. Rev.* 319 (1993).

Perritt, Henry H., Jr., "Metaphors for Understanding Rights and Responsibilities in Network Communities: Print Shops, Barons, Sheriffs, and Bureaucracies," available at http://www.law.vill.edu/chron/articles/metaphor_networks.html (1992).

Perritt, Henry H., Jr., "Tort Liability, the First Amendment, and Equal Access to Electronic Networks," 5 *Harvard J.L. & Tech.* 65 (1992).

Post, David W., "Anarchy, State and the Internet: An Essay on Law-Making in Cyberspace," 1995 *J. Online L. Art.* 3.

Riddle, Michael H., "The Electronic Pamphlet—Computer Bulletin Boards and the Law," available at www.eff.org/pub/EFF/Policy/Legal/bbs_and_law.paper (1990).

Ross, Eileen S., "E-Mail Stalking: Is Adequate Legal Protection Available?" Note, 13 *J. Marshall J. Computer & Info. Law* 405 (1995).

Samuelson, Pamela, "The NII intelectual property report," 37 *Communications of the ACM* 21 (Dec. 1994).

Santisi, Michael J., "*Pres-Kap, Inc. v. System One, Direct Access, Inc.*: Extending the Reach of the Long-Arm Statute Through the Internet?" Note, 13 *J. Marshall J. Computer & Info. Law* 433 (1995).

Sassan, Anthony J., "*Cubby, Inc. v. CompuServe, Inc.*: Comparing Apples to Oranges: The Need for a New Media Classification," Note, 5 *Software L.J.* 821 (1992).

Schlachter, Eric, "Cyberspace, the Free Market, and the Free Marketplace of Ideas: Recognizing Legal Differences in Computer Bulletin Board Functions," Essay, 16 *Hastings Comm/Ent L.J.* 87 (1993), available at www.eff.org/pub/EFF/Policy/Legal/cyberlaw_bbs_free_market.article

Schlachter, Eric, "Electronic Networks and Computer Bulletin Boards: Developing a Legal Regime to Fit the Technology," available at http://starbase.ingress.com/tsw/eric/eric-l.html (1994).

Soma, John T., et al., "Legal Analysis of Electronic Bulletin Board Activities," 7 *W. New Eng. L. Rev.* 571 (1985).

Stevens, George E., and Harold M. Hoffman, "Tort Liability for Defamation by Computer," 6 *Rutgers J. Computers & L.* 91 (1977).

Sunstein, Cass R., "The First Amendment in Cyberspace," 104 *Yale L.J.* 1757 (1995).

"Sysop, User and Programmer Liability: The Constitutionality of Computer Generated Child Pornography," 13 *J. Marshall J. Computer & Info. Law* 481 (1995).

Taviss, Michael L., "Dueling Forums: The Public Forum Doctrine's Failure to Protect the Electronic Forum," Comment, 60 *U. Cin. L. Rev.* 757 (1992).

Tickle, Kelly, "The Vicarious Liability of Electronic Bulletin Board Operators for the Copyright Infringement Occurring on Their Bulletin Boards," Comment, 80 *Iowa L. Rev.* 391 (1995).

Volokh, Eugene, "Cheap Speech and What It Will Do," 104 *Yale L.J.* 1805 (1995).

Yin, Tung, "Post-Modern Printing Presses: Extending Freedom of Press to Protect Information Services," Comment, 8 *High Tech. L.J.* 311 (1993).

The Working Group on Intellectual Property Rights, Preliminary Draft: telnet to iitf.doc.gov [login as Gopher]

Portions of this bibliography were kindly provided by Eric Schlachter of the firm Cooley Godward Castro Huddleson & Tatum, Palo Alto, CA (schlachtere@cooley.com). He notes that it has been impossible to keep up with the explosion of writing in this area.

Cyberspace Law Books

Cavazos, Edward A., and Gavino Morin, *Cyberspace and the Law: Your Rights and Duties in the On-line World*, 1994.

Katsh, M. Ethan, *The Electronic Media and the Transformation of Law,* 1989.

Katsh, M. Ethan, *Law in the Digital World,* 1995.

Rose, Lance, *Netlaw*, 1995.

Rose, Lance, and Jonathan Wallace, *SysLaw,* 2d Ed., 1992.

3 Technology Rights in the International Arena: The Fall of Public Law and the Rise of Private Fiat

The legal system is traditionally thought of as an inhibiting force in the development and marketing of high-technology products. Rules and regulations typically are juxtaposed with profit incentives and with the forces driving development and distribution of products. Much of what we think of as global and trade relations are indeed such an inhibition. Import/export laws, tariffs, and the like all stifle business.

Local businesses, of course, will disagree. Protectionism is designed to assist the parochial industries. These laws protect the nation and its citizens, but they may not advance the state of global economy or of the industry as such.

But there is some law that is indeed designed to protect industries and the technology, to encourage development and advancement without a strong bias towards geography. This law is known conjunctively as *intellectual property* law: the law of trade secrets, patents, trademarks, and copyrights. This law, or something like it, is found in most technologically developed countries.

These protections are moderately important for older industries, such as the automobile, steel, and clothing industries: trademarks, and to some extent trade secrets, can provide significant competitive advantages.

But for rapidly advancing computer technologies, all of these areas—patents, trademarks, trade secrets, and copyrights—are decisive.

A number of factors, all of which relate directly to the computer industries, make these intellectual property protections essential to the well-being of developing technology.

1. Software can be easily reproduced and rapidly modified.

2. Computer hardware is getting smaller and cheaper, and thus, it too, is becoming increasingly reproducible. By this I include faster and less expensive hardware, such as low-cost, extremely fast, high-resolution video hardware, and increasing power, more instructions and floating operations per second, per dollar.
3. We have better, cheaper, more comprehensive, and widespread software development standards, specifically including programming languages such as C++.

These factors simultaneously (1) represent powerful, useful advances in the technology, and (2) radically open up the possibilities for unfair competition—more people in more countries can steal technology than ever before.

Where intellectual property law is absent, or practically absent because it is unenforced, we find a powerful disincentive to business. The conditions in Spain, Greece, and China are abysmal for software companies, because there is in general practice no copyright law in those countries. Thus, there is no great rush towards increasing our software investments there, even though these countries may otherwise provide excellent opportunities.

But there is a spectrum of protection here. There is such a thing as too much protection for the inventions of the intellect and imagination, and the results can be as catastrophic as too little protection. This is because protection of one product always means barring another newer one that is, in some respect, too close to the protected product. For example, patent rights block a subsequent invention that essentially does the same thing in essentially the same way. Copyright stops later authors from incorporating earlier works into their new products, and new trademarks can be enjoined if they are confusingly similar to more senior marks. It takes a delicate sense of judgment to compare the older and newer works to discern infringement; that judgment, exercised by juries and judges unfamiliar with technology, is not predictable.

Thus, an enforced set of balanced intellectual property laws is quite as important for the local industry as it is for foreigners. For example, weak enforcement of software copyright laws in India not only discourages U.S. companies, but will inhibit the development of the indigenous software industry [1].[1]

---

1. Note that, however, India's software export industry rose from about $10 million in 1985 to a projected $350 million for 1993 [2].

THE GLOBAL CONTENT

Market forces have conspired to create a global technological environment. Computer-based industries are in some respects simply a function of that broader evolution.

This is the result of three important trends: (1) the global source of components; (2) the global identity of the players; and (3) the global nature of telecommunications, data, and the tools (such as virtual reality) used to manage the information.

The first two trends—products and companies—explain a traditional development of international intellectual property development; I shall address those first. I will reserve the last—ubiquitous data and communications—until the end.

Global Products

Products are assembled from components manufactured throughout the world. Computers are a prime example: they are stuffed with items from Malaysia, South Korea, Taiwan, and Singapore. This is true for software as well, although the sources are generally not as obvious: chunks of code are written in India,[2] while data is imputed and assembled in the "data barns" of the Caribbean and Mexico for low-cost data entry and telemarketing [4,5].

Global Companies

While every firm is nominally a citizen of a given country, the players are in fact international entities, whether alone or in conjunction with their strategic partners.

Most obviously, IBM, Hewlett-Packard, Phillips, and such are in fact international entities. These corporations will shift facilities and employees as necessary on a global scale. These companies are stateless, without national loyalty [6]. Many of these companies are, in fact, owned by stockholders from various parts of the world: British Air is 25% owned by Americans; England's W Industries used to control America's Spectrum Holobyte. Fujitsu owns 80% of the U.K. computer manufacturer ICL. The Japanese subsidiary of the U.S. firm Texas Instruments moved its 4-MB DRAM manufacturing facilities to Taiwan and Singapore [7]. Cultural

2. See generally [3]. Still, the United States retains the lead here. U.S. firms hold about 75% of the global market for prepackaged software and approximately 60% of the world market for software and related services. In 1991, foreign sales of U.S. prepackaged software vendors totaled over $19.7 billion.

Technologies, a division of Fujitsu OSSI located in northern California, was working on a graphical interactive networked entertainment (cyberspace) project, translating a Fujitsu network product for the U.S. telecommunications environment.

Other firms, while domestic, have strong strategic partners abroad, used to assemble and manufacture products. General Electric's microwaves are from South Korea, Chrysler sells Mitsubishi cars, and so on. Software giants ASCII and Microsoft sell Windows NT in Japan, and Sun Microsystems has an investment in Dublin-based software developer Iona.

Many companies deal through wholly owned foreign subsidiaries. For example, Sega, Nintendo, and scores of other computer companies own corporate entities in various countries.

The actors of industry are, in fact, international entities selling global products.

DISPARATE PROTECTION OF GLOBAL INTELLECTUAL PROPERTY RIGHTS

While recent trade accords suggest hope for harmonization, there is a wide diversity in practice. The fact is that the global creature of advanced computing technology is governed by a myriad of inconsistent laws. Here are a few examples:

- Nations are divided on the extent to which the "look and feel" of software is protectable by copyright. Indeed, nations are divided as to whether copyright protection extends to software, although that is the clear tendency. They are also divided on the extent to which doctrines of "fair use" protect certain sorts of copying and decompilation.
- Nations have differing laws on the extent to which, if at all, software is the proper subject of patents.
- Patents generally must be prosecuted in every country of interest. Although there is some ability to have single prosecutions for multiple countries,[3] countries have different rules on, for example, the effect—which can be fatal—of prefiling disclosure on the validity of the patent application.
- The United States has first-to-invent rules for patent priority; most other nations, including Japan, have a first-to-file rule.

---

3. Fourteen countries are party to the European Patent Convention, and 47 are party to the Patent Cooperation Treaty.

- Some terms, marks, and symbols can be registered as trademarks in some countries, but not others, which undermines an entity's desired global corporate identity.
- It is expensive to register trademarks in all countries of interest to a client; with a few exceptions, such as the single filing for the European Community, one is required to have separate national filings.
- Trade secrets in Japan are very difficult to enforce, and, unlike in countries such as the United States, such secrets must be publicly disclosed to be litigated.
- Countries have vastly different laws on a variety of import and internal distribution regulations and practices, including grey-market imports. These occasional, informal, invisible practices can be quite as important to the viability of one's product in another country as the formal laws on, for example, intellectual property.
- Various countries make illegal the dissemination of indecent materials, such as materials "unsuitable" for minors involving drugs, sex, and violence. The United States passed a statute to similar effect but that was ruled unconstitutional; most nations do not have a similar constitutional structure.

What of late developments, such as the recently concluded Uruguay Round of the GATT negotiations [8]? Limited progress was made. The signatories to GATT adopted much of agreements from the Berne convention and expressly extended copyright to software and database compilations. GATT also includes obligations under the Paris patent convention, requiring 20-year patent protection for most inventions. Trade secrets must be protected, and criminal penalties for willful trademark counterfeiting and copyright piracy must be enacted. The countries must also put into place procedures concerning the use of evidence, injunctions, damages, seizure, and destruction of infringing goods.

Transition periods range from one year to eleven years for the so-called less developed countries. Even then, of course, it is impossible to predict the still-wide range of enforcement mechanisms that may be created, and there will remain many countries in which, without an effective court system, GATT is no more than a few words on the statute books.

GLOBAL PRODUCTS AND PAROCHIAL LAW

The computer industry, and especially networked interactive telecommunications aspects of computer technology, is itself a global phenomena in a manner quite distinct from examples I have previously provided (such as the multinational corporation and chips made in one country for use in another). And this, in turn, generates legal issues quite distinct from the traditional, multilateral, intellectual property matters I have previously alluded to.

It is important to realize that computers and telecommunications are becoming the trade routes for other products, as well as creating new computer-based products [9]. These new trade routes or vectors, coupled with the new products and services, have combined to further the expansion of business and economic entities far beyond national borders—making national borders in some instances superfluous. We are creating shared worlds, which transcend the physical worlds of nations and boundaries.

Let me briefly trace this progression. Traditionally, the law that governs property disputes is the law where the property physically stands. Intellectual property can be leased, assigned, bought, and sold like a gold piece or a pig. As economies expand and increasingly rely on trade—or, better, as we increasingly realize the interdependencies—nations begin to compete globally; rather than rely on trade barriers, companies compete on their merits with better products, cheaper products, or faster response times.[4] Consequently, we have a diminution in the importance of the nation state; borders become more permeable. Products made within a country's borders are comprised of components from without the borders, including key technologies required for its military defense [11].

Economic zones take on great importance [12]. Global markets for goods and services, with modern marketing techniques and technologies, shape the economies of region states such as Hong Kong and Southern China; San Diego and Tiajuana, Mexico; Singapore and its neighboring Indonesian islands. Some countries, such as the United States, are made up of different region states.

Now let us extrapolate. We do not exaggerate much when we say that, in principle, information is now instantly available throughout the developed countries, if not the globe [13]. Information, as such, is a commodity, a new property, and it has value; *information manipulation* is an increasingly valued service.

This suggests products and services extant in transnational cyberspace. These are global products: they persist, diffused as a network is a diffused entity, diffused as the data flowing to databases around the world.

---

4. Even the Japanese, who generally may have thought their economy immune from outside pressures, know differently. And that knowledge is assisting Japanese business in becoming a stronger global competitor [10].

I quote an article on an object-oriented approach to operating systems design. The author assumes—rightly, I believe—radically distributed systems [14]:

> The seamless nature of object systems will radically alter the way we think about *where* our data is. Data will be encapsulated in objects that will in some cases be able to roam to where they are most needed. We are in the habit of thinking that a document is simply stored on a particular hard disk. Distributed object systems will ask us to surrender that comfortable certainty in exchange for the power and flexibility of location-transparent storage.

In this context we find *true* globalization of data, and consequently of the technology systems that will be used to handle the data. It is impossible to distinguish governing parochial law, because there is an intractable uncertainty in choosing a nation's law to apply. For example:

- Electronic funds transfers: Whose laws on accounting should be enforced? Whose regulations on reporting?
- What law applies to regulations for stock and other international financial markets?
- Eavesdropping, wire or data interception: Whose laws on the propriety of search and seizures? On data accumulation? On spying?
- Satellite distribution of films and other entertainment including linked interactive virtual reality entertainment: Whose law on copyright applies?
- Whose laws restricting distribution of, say, pornography or state secrets, applies?
- Posting of cryptography software, such as Pretty Good Privacy, on the Internet, which may violate U.S. or other countries' import/export laws. It is legal to use the software within the United States, but it is a crime in France.
- Virtual reality/telepresence medical procedures: imagine a patient in one country and doctors in others. Whose law on professional negligence applies? How about privileges regarding good samaritan immunity?
- Global "publication"/availability of news and information: whose law on copyright and what mechanism to collect royalties will apply [15]? Whose law on libel? When a Swedish company arguably libels a California resident in a globally distributed story, the U.S. courts may not have the jurisdiction to try the case [16].

- Whose law on the electronic disclosure of trade secrets should be used?
- What law does one apply on the intentional infliction of emotional distress? How about broken promises of marriage—enforceable in some jurisdictions, but not in others?

These examples should be enough to emphasize the situation: We are confronted with a cacophony of rules and regulations, wholly incommensurate with the global entity known as dataspace (or cyberspace) and its associated networked computer systems.

The classic notion of law, of course, presupposes the direct action of the legislating sovereign nation; the individual nations have been, and are still, the source of law. Recall that GATT's intellectual property provisions are not international law in the broad sense—rather, GATT requires member nations to come up with compatible *national* law.

To be sure, there will continue to be a role for such national law and multinational treaties. Companies still need their manufacturing plants somewhere. Software developers and academics still need a place to work, local rules on secrecy, and some confidence that at least the host nation will provide truly enforceable copyright protection, for example. Governments that wish to attract these development efforts, and to share in the power that they afford, will have to "play ball" and seriously enforce intellectual property rules.

Nation states and their laws still have a place in facilitating advanced technology products. These countries *must* enact reasonable, balanced, and enforced laws on the protection of intellectual property. If they do not take these steps, international sanctions will not be the highest price paid by recalcitrant nations. Instead, they will be left to wither on the vine, bypassed by computer technology and by the rest of the world's business that is managed by that technology.

But if geographically bound sovereign nations are increasingly irrelevant to the conduct of global trade in general, and to global computer phenomena in particular, does that not suggest a new age of lawlessness?

To some extent, it does. There is an aura of futility in the attempts of some countries to regulate what, by its nature, cannot be regulated. The efforts of the U.S. government to bar export of sophisticated encryption software are absurd, just as would be efforts to stop the import of data from the Caribbean and Mexico, just as it is unreasonable and ineffective for the United Kingdom to censor books widely published abroad. And when the United States has a true information highway in place feeding millions of homes, it may not be possible to effectively regulate, for example, the "import" of grey-market home-entertainment video games.

Perhaps we are discovering the limits of powers and periphery of the law. It may well be that trade regulations *cannot* help much with dataspace; it may well be that copyright law *cannot* protect against the use of counterfeit chunks of code.

Recall that the pressures of economic technology transfer are towards the less developed countries—from the United States to Mexico, Japan towards Thailand; from Taiwan, perhaps, towards Vietnam. And those less developed nations are precisely where all of the laws we have discussed are the *weakest*.[5] International corporations will tend to move to countries where laws and regulations permit the greatest degree of freedom from public law.

If we cannot rely on the public law, then what is left? What happens when countries are increasingly powerless to control rapidly developing and expanding technology? The answer is that the technology will be controlled by the international players who control the means of creation and distribution of the product. We are, and will be, governed by private contracts and regulations put in place by the providers. Nintendo, Sony, and Sega control the content of millions of games, regardless of who writes, publishes, or plays them; Motorola may control the use of satellite communications; banks and credit card companies can and do control the use made of trillions of bytes of personal data; perhaps Microsoft, Netscape, and Sun Microsystems will control standards and compatibility of the computing systems. And encrypted data and privately locked software will fill the dataspace.

This is not law in the way we usually think of it. These are private arrangements, and private fiat, required for the stability and predictability that the law in a rapidly changing and global environment cannot give. The lawyers must then look to private dispute resolution, and away from the public litigation systems of nations; we will worry less about patent law in Thailand and more about cross-cultural negotiation, mediation, and arbitration techniques.

Business will still cooperate with the laws of every country in which it desires to protect the works of its labor and imagination. But, ultimately, that public law may provide scant protection: private rules and private fiat will govern as the nations turn infirm.

References

[1] Weisband, Suzanne, and Seymour Goodman, "Subduing Software Pirates," *Technology Review*, Massachusetts Institute of Technology, Oct. 1993, p. 30.

5. See [17,18].

[2] *The New York Times* (Dec. 29, 1993).
[3] Gargin, Edward, "India Among the Leaders in Software for Computers," *The New York Times* 1 (Dec. 29, 1993).
[4] Swardson, Roger, "Greetings From the Electronic Plantation," *Utne Reader* 91 (March/April 1993).
[5] McKenzie, Richard, "Electronic Immigrants Are Taking Jobs From Americans," *San Francisco Daily Journal* 4 (Dec. 15, 1993).
[6] Greider, William, *Who Will Tell the People,* New York, 1992, p. 394.
[7] *Software Opportunities in Japan,* Vol. 2, Dec. 1993, p. 3.
[8] Agreement on Trade Related Aspects of Intellectual Property Rights, Including Trade in Counterfeit Goods (TRIPS agreement), contained in the Draft Final Act Embodying the Results of the Uruguay Round of Multilateral Trade Negotiations (GATT) and Chapter 17 of the North American Free Trade Agreement (NAFTA).
[9] Reimers, Neils, "New Economy Requires New Management," *Les Nouvelles* 154 (Dec. 1993).
[10] Drucker, Peter F., "The End of Japan, Inc.?" *Foreign Affairs* 10 (Spring 1993).
[11] Moran, Theodore H., "International Economics and National Security," *Foreign Affairs* 74 (Winter 1990–91).
[12] Ohmae, Kenichi, "The Rise of The Region State," *Foreign Affairs* 78 (Spring 1993).
[13] Woolley, Benjamin, *Virtual Worlds*, Cambridge, MA: Blackwell, 1992, p. 124.
[14] Peter, Wayner, "Objects on the March," *Byte* 139 (Jan. 1994).
[15] Moshell, J. Michael, and Charles E. Hughes, "Shared Virtual Worlds for Education," *Virtual Reality World* 63, 69 (Jan./Feb. 1994).
[16] *Core-Vert Corp. v. Nobel Industries, A.G.*, (U.S. Court of Appeals for the Ninth Circuit, December 16, 1993).
[17] Nimmer, R., *The Law of Computer Technology*, ¶ 5.02, 3 (Rev. Ed.), 1992.
[18] Greider, William, *Who Will Tell the People*, New York, 1992, p. 399.

4 The Uneasy Treaty of Technology and Law: A Summary of Legal Issues for the Virtual Reality Industry

When you are gone, Reason gone with you,
Then Fantasie is Queene and Soul, and all

—John Donne, Elegie X, *The Dreame*

This chapter describes the interplay of two systems: the law and computer technology, specifically virtual reality (VR) technology.

Poised as it is at the intersection of a variety of industries such as entertainment and consumer electronics, VR inherits all their legal issues as well. More directly, the legal system treats VR as a subset of computer applications, and the law reacts to VR as it does to any rapidly developing technology. These aspects of the technology carry a classic assortment of legal baggage, outlined below. Perhaps more intriguing, however, are those special issues that will arise as the true power of VR technology erupts: as the yield of mass hallucination, and figments of computer imagination, converge to design and control real-world events.

THE INDUSTRY

The budding virtual reality industry is a hybrid—a mixture of applications, hardware, and strategic directions. The community includes the small "home brew" developer and massively parallel supercomputing systems operated for the Department of Defense. The VR label has been applied to desktop computer flight simulators and multimillion-dollar Las Vegas Luxor and Disneyland Star Tours space rides—3D computer-assisted design (CAD) programs that display highly complex molecules—to the Nintendo monochromatic video game headset. The label applies to gloved, helmeted engineers reaching into cyberspace to manipulate fake space objects in the design of real, honest-to-God nuclear submarines. Bytes and pieces of virtual reality lurk in our use of cash machines, in the way we meet on electronic bulletin boards, and in our use of voicemail.

VR: THE ENTERTAINMENT LAUNCH

Entertainment had led the way in VR development: witness the consumer home video game headsets from Sega slotted for 1994, Nintendo's headset, and those released by VictorMaxx. Stereo vision and 3D sound software designed to drive the hardware has been available for over a year. Witness the arcade and mall installations: the BattleTech, Virtual Worlds, FighterTown, and proposed Star Trek and other VR environments and "rides." Together with the potential product liability issues described earlier, these production and marketing ventures involve a raft of legal issues not formerly associated with the computer industry. These include contracts with actors, guilds, directors, artists, musicians, as well as communications and regulatory law, as cable and other delivery systems are used to mediate interactive VR experiences.[1]

THE COMPUTER

VR applications have in common the computer. And the issues facing the VR community are generally no different from those facing the computer industry in general. The legal issues are deep and wide: there is no immunity here. Any company can have contract problems, environmental regulation issues, or labor employment disputes. All computer corporations have to deal with copyright issues: they want to protect their ownership of software, text, or art, and need to avoid stealing material from others. They all run into trademark problems from time to time, as they seek to have the consumer associate their products with a word or design. Trade secret disputes erupt periodically: secret strategic plans reside in the brains of high executives being courted by a competitor; or a technician or programmer leaves with a customer list or blueprints—and starts up her own company.

These can be vicious, nasty problems, and they can destroy a company. But, again, they are not the special province of VR. These problems

1. Now, business software is on the rise. Leading companies in the defense, aerospace, and automotive industries are utilizing VR technology for design and general problem solving tools [1].

are not even the unique province of companies in the computer software or hardware business: satellite fabricators, engineering corporations, and toy manufactures confront the same legal issues. And as computers reach out into every vein of the society and become the essential building block of the new infrastructure, computer-related issues become the issues of everyone.

To be sure, the rapid growth and ubiquitous technical aspects of the computer industry do lead to special difficulties. Judges just getting comfortable with 1960s civil rights laws and 1980s employment law have a hard time with 1990s software copyright issues. Statutes written in the last century are used to sort through computer ownership matters, and ancient analogies are used to gauge whether sophisticated encryption code barred from physical export can be posted on electronic bulletin boards.

There are inherent delays in the judicial system—it takes literally years in many California courts to get to trial—but the effect of that lag time is aggravated in the computer industry. Courts construe terms and words, adjust rights and responsibilities, and dictate the outer limits of acceptable behavior. Companies try to play by the rules, but when the industry's terms, words, products, and applications have a shorter life than the average lawsuit, the legal system is left in the dust—and the industry is left twisting in the wind.

VR: RAPIDLY DEVELOPING TECHNOLOGY

But if the rapidly developing computer industry tests the legal system, then VR—the cutting edge of the computer industry—may break it. As a hybrid domain, VR pushes the envelope in a variety of technologies: large computing machines and complex architectures, new lightweight input and output devices, and multithreaded and multiprocessor programming in the computer industry, as well as entertainment, medicine, engineering, physics, and other sciences. As it integrates and synchronizes these fields, VR becomes more than its parts.

If the number of papers, conferences, and articles about VR are an indication, its growth is explosive. The technologies for a fully immersive, complete VR experience are already on the market; only issues of cost and size keep them from the hands of consumers and small business. The VR Technology Task Group, part of President Clinton's High Performance Computing, Communications and Information Technologies Committee, has determined that VR is probably the nation's most important technology of the decade, and global spending for VR technology was forecast to exceed half a billion dollars by 1997 [2].

The prognosis is for a great deal of new technology in a short period of time. Uncertainty and misunderstanding are classic responses. New technologies lend themselves not only to hype but also to phobia: unreasoning hope and unreasoning fear. Everyone is an instant expert, physical and mental injuries can be supposed—and the lawsuits are filed. Where direct proof of causation of injury is unavailable, experts and lawyers uses studies and statistics, which can show almost anything. After all, anything *is* possible. Will VR eyewear damage the brain or the retina? Fry the corpus callosum? Cause epilepsy? Experts and lawyers have said worse: recall the litigation over electromagnetic fields from power lines and PC monitors, allegations of brain cancer from cellular phones, fears of genetically engineered tomatoes, and suggestions of murderous violence as a result of home video games. With new technology, one never knows.

That is a bird's-eye view: VR as a new technology, as the intersection of multiple fields, heir to the legal problems of them all.

VR: UNIQUE PROBLEMS

But plunging down from the bird's-eye view, there are as well peculiar and specific legal issues for the VR industry. The two major branches of mature VR applications are those that (1) control real-world systems and (2) are used for analysis. This recalls the division suggested by Shoshana Zuboff: computers (1) automate and (2) informate [3]. In each case, VR is used to handle colossal amounts of data—and therein lies the rub.

The first category comprises real-time command, control, and communications systems. Examples include air traffic control, telepresence technology to control remote robots and drones (in space, underwater, or in other hostile environments) as well as miniature medical devices within the body, VR environments for those controlling power stations, and advanced fighter aircraft and associated weapons systems.

The second category consists of design and analytical tools. On the desktop, we have VR interfaces for the collection and analysis of stock market and other complex financial data—one swoops and glides among colored towers and undulating hills, representing market volume, prices, and moving averages. Boeing employed advanced computer-assisted design–computer-assisted modulation (CAD-CAM) techniques, virtually virtual reality, for the design of its 777 commercial jetliner, and new consortiums in the United States and Europe have recently been formed to use VR technology for the interactive design of portions of nuclear submarines, other shipbuilding, and further large, complex projects.

As these examples suggest, the true purpose and strength of VR is the management of enormous amounts of complex data. These complex applications are the focus of large sums of research money, and it is here that we find legal issues peculiar to the VR industry.

To place these issues in perspective, one must recall the purpose and effect of the technology. VR is aimed at what I term the "zero user interface." It is focused on the creation of an illusion of immersion in the computer-generated reality. Ideally, VR does away with the abstractions of data such as numbers, and it does away with the machine itself: keyboards, other peripheral devices, are erased. This sense of illusion generation is suggested by some of the names in the industry (Fakespace, Illusion Inc., Visions of Reality), as well as the generic title "virtual reality" itself.

The point, after all, is direct sensory input, as direct and natural as the connection we have with the "real" physical world. In short, the point is to *fool the user:* "consciousness never deceives" [4].

At the same time, VR applications are nothing but models of the real world, and as models, they are abstractions. VR programs filter out pieces of the truth and they mutate the rest; they alter the segments that *are* presented to the user.

An example will help. A pilot of an advanced tactical fighter may wear a VR helmet that presents a model of the terrain and targets in the environment. The representation is an abstraction: as the Latin root of the word suggests, this means that certain things have been *taken away.* Trees and bushes, rivers, perhaps buildings, are invisible, while mountain crests and tanks are visible and represented in easily identifiable colors—say, blue for the good guys and red for the targets. The program also *mutates* the data that is shown: blocks instead of tanks, perhaps color for target type, or sound intensity or frequency to represent distance.

VR achieves its capability to manage large amounts of data by first filtering out "irrelevant" information and then enlisting other senses for the input of the presented information. Data is, in effect, spread out so that the senses of hearing, touch, and visual depth perception are all engaged. The VR presentation is, ideally, overwhelming, and it leaves no room for any other input of any other kind from the "real world"—for that would destroy the illusion.

The result is presented to the user in the most convincing way, but the result is, simultaneously, (1) a lie and (2) impossible to check.

To have a feel for the legal significance of this discussion, one must recall a few other characteristics of software: First, it is not possible to completely debug a complex program. Second (and this may be just a corollary of the first observation), it is not possible to completely predict the results of tightly coupled man/machine links. Unpredictable oscillations have been noted in a variety of circumstances, including the "computerized" or "fly-by-wire" systems in the early space shuttles, the European

Airbus, and the Chernobyl nuclear power plant, in which human attempts to ameliorate a crisis (by feeding input to the machine) instead aggravated the situation.[2]

Not to put too fine a point on it, the interest of the legal system should now be apparent. These systems are inherently complex. They are designed to manage large quantities of valuable data or expensive machinery—and at the same time may co-opt their human controllers, recasting those controllers into becoming *part of the system.* It can be argued (as a lawyer might say) that these circumstances increase the possibility of catastrophic failure. Where property damage, injury, and frustrated expectations abound, there lawyers be.

CONCLUSION: A TALE OF TWO CODES

There is a deep contrast between the approach of rapidly developing technology and the highly conservative instincts of the law. One looks forward to risks, speculation, and experiment; the other looks back to precedent for legitimacy and analogies. New technology, represented by VR in one of its most exciting incarnations, cannot wait for certainty or complete assurance in all its effects. But by its nature, the law prizes the "reasonable" and "prudent" person, and often punishes the risk taker.

Which will bend? In the short term, obviously new technology companies must comply with the law, even its unpredictable and confused instances. Checks and balances will be created for inherently dangerous applications. Warning labels will be used; contracts will be negotiated very carefully, with an eye to an unimagined future.

But in the long term, as computers become part of, and disappear into, the environment, I suggest that their effects will be treated *as* environmental, as ubiquitous as weather, and as necessary as air. Some risk will become part of the backdrop, just as today we assume the risk posed by cars and certain levels of potential carcinogens. Such is the extrapolated effect of the zero user interface.

To be sure, we will draw lines to attack the unacceptable risk. But those lines, too, are unpredictable, and will shift across the social landscape and over time. Law and technology will continue their uneasy embrace, intimate and suspicious.

---

2. This problem was bought to popular attention in 1972 when Michael Crichton published *The Terminal Man*. There he described a man/machine interface out of control as a result of positive progression cycle: input from one side of the interface (the implanted computer) generated input for the other side (the human brain), and vice versa, without a check from the "outside world."

References

[1] "From the Editor," *Internet World* 10 (Dec. 1996).
[2] *PC Week* 147 (Oct. 4, 1994).
[3] Zuboff, Shoshana, *In the Age of the Smart Machine*, New York: Basic Books, 1984.
[4] Hume, David, *Enquires*, Selbey-Bigge (ed.), Oxford: Oxford University Press, 1902, p. 66.

Data Morphing: Ownership, Copyright, and Creation

5

INTRODUCTION

I discuss copyright not because it is a tool of choice in the current software wars, but because copyright promises to be increasingly applied to the disputes erupting in the virtual worlds of the future.

Copyright is cheap—it costs nothing to secure and only a few dollars to register with the Library of Congress.

Copyright protection is broad—it bars anyone but the author from using the work, displaying it, making derivative works, making copies, or undertaking performances of it.

Copyright protection is powerful—it can block the distribution of a "substantially similar" product, and the scope of that bar can be very wide, depending on a range of imponderable variables. For that reason, copyright can be dangerous: designed by the writers of our Constitution to be a powerful incentive to innovation and creativity, it can stifle competition, halt valid alternatives, and punish novel improvements. And because copyright is a *legal* doctrine, and is implemented by the *legal* process, it suffers from the disease afflicting all doctrines of the costly, overburdened, and overcomplicated legal system: copyright, and expensive copyright suits, can put a company out of business and springboard another into preeminence.

In short, copyright is decisive: it is *the* means by which rights are allocated; it defines the relevant meaning of "property."

In this chapter, I will first outline a few—although not all—basic propositions of copyright law. I will then discuss current copyright problems having to do with the digitizing and mutating of data—issues that will be familiar to those wondering just how much one can get away with in the creation of multimedia products.[1] Finally, I will make a few assumptions on the structure of broadband, interactive, multisensorial virtual reality, and then outline the clash—the irreducible opposition—between the realities of the new technology and the fundamental ideas of the old law.

BASIC PROPOSITIONS

The fundamental concepts surrounding copyright can be summarized as follows:

> Copyright is the right of a creator that automatically inheres in every work of creation that is fixed in a tangible medium. Copyright never protects the *ideas* in a work, only the expression itself. The owner of the property can stop everyone else from using, copying, displaying, or in any way exploiting his expression, unless the new use is a so-called "fair use." A copyright owner can get a court order blocking someone else's exploitation of his work, an exploitation that is tantamount to theft.

Most importantly for our purposes, the copyright extends to all works which are "derivative"—and all works which copy enough of the original work so as to be "substantially similar."

If this sounds a little loose to you, you heard correctly. There are fundamental flaws in the notion of copyright, and these flaws are brought to the front by the application of the law to new technology. The legal evaluation of current technology is problematic; its application to future developments is unmanageable.

The basic concepts, then:

- Copyright protects the actual creative expression,[2] not the ideas in the expression: it protects the *Mona Lisa*, not the idea of an enig-

---

1. I do not directly discuss the critical topic of computer software copyright. There is insufficient space here to consider the rapidly shifting and (sometimes) developing law. (For an update, see [1].) More importantly, while the VR community will always have concerns about the scope of computer software copyright, those concerns will not differ significantly from those of the computer industry generally. Those involved in the creation of VR products, on the other hand, I think will have a *specific* interest in the copyright problems associated with the digitalization and transmutation of *data*, which is the general topic of this chapter.
2. Most any form of expression that can be fixed in some fashion can be the subject of copyright, including all forms of writing, including computer code (object, source, and

matic smiling Italian Renaissance lady before a distant background. It protects your drawing of an orange, but does not stop anyone else from drawing his own orange.

- Copyright presumes a creator, an author, the modeler of clay, the weaver of notes, the painter, the poet, the architect designing the lines of a building, or the software author writing line after line of code.
- Copyright requires the created materials to be fixed and tangible.[3]
- Copyright bars any use, derivative material, or other exploitation of the work, including any new work that reproduces enough to be "substantially similar" to the original.
- "Fair use" is a defense: that is, it protects use which is for research, for educational purposes, for comment on, or for parody of the original.[4]

BASIC PROBLEMS

Authorship

The author is the owner; without the author, there is no owner, and because copyright is a species of owned property, copyright makes no sense without an author.

Sometimes it is easy to determine the author: we credit the scrivener and forget about his or her influences. Shakespeare and not his family and friends wrote *King Lear*; T. S. Eliot, not his hardworking editor and fellow poet Ezra Pound, wrote *The Waste Land*; Picasso, and not his colleagues with whom he spoke on war and politics, whose advice he took, and whose works influenced him, painted *Guernica*.

Creativity—a modest, small amount of originality—is required to secure protection. But everyone who contributes to the final work may be treated as a joint author, with full rights in the completed work.[5] And so

---

micro code), audiovisual displays, sounds, and all graphical, architectural, broadcast, and sculptural works.

3. Fixed and tangible embodiment includes the news broadcast as it issues out into the airwaves or the home video game screen as animations flutter across the display. The point is that the work can be viewed or read by a human, with or without the assistance of a machine (such as a computer).

4. See generally, *Metro-Goldwyn-Mayer v. Showcase Atlanta Co-Op Productions, Inc.*, 479 F.Supp. 351 (N.D.Ga. 1979); *Salinger v. Random House, Inc.* 811 F.2d 90 (2d Cir. 1987).

5. Joint authors are automatically established when two or more persons intend that their contributions be merged into an inseparable—or interdependent—parts of a unitary whole. 17 U.S.C. § 201(a).

the troubles begin: 1% of the contribution may provide a joint author with a full 50% of the rights to direct what shall be done with the work—and 50% of the profits.[6]

The authorship requirement also means that no one can copyright *facts* and *events*.[7] The realities of the world are in the public domain, including historical, geographical, and other scientific and sensory fact. These belong to everyone. And when the expression of the fact is inseparable from the fact itself—such as when a map depicts a pipeline in the only way a pipeline can be depicted on a map—then the entire expression—map and all—merges with the idea, and there is no copyright protection at all.

Even more interesting for our purposes is the doctrine of "scènes à faire."[8] Under this principle, entire conglomerates of description, phrases, views, sounds, and pictures are in the public domain when they constitute the classic or traditional method of representing something. Plots, incidents, characters, or settings that are traditionally associated with a situation are not attributable to the creativity of the author, and so are not protected. Examples include portraying "drunks, prostitutes and vermin" in depictions of police life in the South Bronx[9] or a broken-hearted lover listening to juke box country music in a bar.[10] Clichés, in short, belong to us all.

Fair Use

Once a work is done and published—sent out into the world—what are the permissible uses?

The "fair use" doctrine allows use of another work for criticism, reporting, comment, teaching, scholarship, research, and parody—a form of criticism. Fair use can be a powerful defense, but it is often defeated when the new user makes money, and almost always fails when the new use takes away sales from the original owner. This last generalization suggests an important corollary: When a creation is shifted from one medium to

---

6. See also the notion of a "work for hire." Under that principle, the employer is owner when the work is made by the employee in course and scope of his duties. See generally [2].

7. *Georgia Television v. TV News Clips of Atlanta*, 718 F. Supp. 939 (N.D. Ga. 1989); *Applied Innovations v. Regents of The University of Minnesota*, 876 F.2d 626 (8th Cir. 1989); *Feder v. Videotrip Corp.*, 697 F.Supp. 1165 (D. Colo. 1988).

8. Literally, "scenes which must be done."

9. *Walker v. Time Life Film, Inc.*, 784 F.2d 44 (2d Cir.), *cert. denied*, 476 U.S. 1159 (1986).

10. *Black v. Gosdin*, 740 F.Supp 1288 (M.D. Tenn. 1990).

another, and the new work does not compete for the same dollars as the old work, there is somewhat *higher chance* that fair use will be found.

But the notion of a "derivative work" cuts across all media, and includes all transformations and adaptations.[11] Such transformations are theft when the new work "substantially" borrows from the copyrighted work's substance and expression. For example, the original owner retains all control over translations, fictionalizations, motion-picture versions, sound versions, art reproductions, abridgements and condensations, adaptations, or any other transformation. Every single one of these can be the subject of separate copyright[12] and the rights to all are owned by the original creator.

Idea/Expression

Copyright will not protect the idea, only the expression. The problem is that no one really understands how to apply this in a given situation. Without infringing Edward Albee's rights in *Who's Afraid of Virginia Woolf* one could write a dramatic work about a feuding couple who play mind games to destroy each other, but if you also use the idea of a college locale, provide elements of the characters' interaction, throw in another innocent couple as Albee did, and start paraphrasing the dialogue, infringement will at some point be found. This is because the notion of protectible "expression" goes beyond the literal text. Copying a work's arrangement, its structure—its look and feel—can be held to be theft of the "expression." If this were not so, minor but global modifications to a work would confer immunity upon the thief. On the other hand, trivial congruences between two works will not establish infringement.

How much and what *sort* of copying, constitutes copyright infringement? The test is "substantial similarity." If the expressions in the original and new works are "substantially similar," then there is infringement.[13]

Similarity of Expression: Key to Infringement

But what is substantial? What does it mean for something to be "similar" to another? These questions, which lie at the very heart of copyright law,

11. 17 U.S.C. § 101. Photographs may infringe ballet choreography, *Horgan v. Macmillan*, 787 F.2d 157 (2d Cir. 1986). Motion pictures can infringe a book, *Stewart v. Abend*, 495 U.S. 207 (1990).

12. See also the notion of a *collective work*, a compilation or assembly of preexisting works, including the organization and arrangement of items which themselves independently may be the subject of copyright.

13. See, for example, *Worth v. Selchow & Richter Co.*, 827 F.2d 569 (9th Cir. 1987).

are fundamentally unanswerable in the abstract. Cases go every which way on these key problems. Some look to the locale, setting, characters, situations, and relationship of elements in the two works being compared, such as with two dramatic pieces.[14] The issue has been phrased as whether a substantial portion of the original work is used; however, this doesn't help either because using a small, but highly important, piece of the original work can be an infringement.[15] A "value" judgment is required to decide whether the stuff, the items, the elements, of the original work used in the new work were qualitatively "substantial."[16] Under certain circumstances, a few notes, a few sentences, a few cropped parts of a photograph, a few lines of code—these all *could* be substantial parts of an original work, off limits to the rest of the universe.

Of course, before one thing is thought of as substantially similar to another, one first must have decided that the two items were similar to each other.

Whether an element or object is similar to another is a matter of context and purpose: two things are, or are not, similar depending on the situation. Take three glasses of liquid: the first two are clear, and the third, red. You would think that the first two glasses are more similar, have more in common, than either does with the third; unless the first and third glasses contain water (one with red food dye) and the middle glass contains clear hydrochloric acid—and I am thirsty.[17]

Consider, if you will, two fonts—fonts can be copyrighted[18]—and the accusation that one is substantially similar to another. Is the small "a"

---

14. *Metro-Goldwyn-Mayer,* supra.

15. 3 M. Nimmer & D. Nimmer, *Nimmer on Copyright* § 13.03[A] (1992). Cases are cited there in which, for example, a *qualitatively* important piece of the original work was used, but that piece constituted only 0.8% of the original work: the infringement was "substantial" nevertheless. *Id.* at p. 13–48 and note 99.

16. Nimmer, op. cit.

17. Perhaps the single most important article written on similarity is by Nelson Goodman, who taught philosophy at Harvard University. Professor Goodman points out that noting "similarity" between any two things is gratuitous, as all things have practically an infinite number of properties in common. For those readers enamored of basic set theory, Goodman provides a simple proof buttressing his warnings against the use of "similarity" and shows that [3]: (1) Where there are n things in the universe, each two things have in common exactly 2^{n-2} properties, out of total properties equaling 2^{n-1}; and (2) Where there is an infinite number of things in the universe, all these numbers, including the number of properties any two things have in common [and thus the ways in which the two things are "similar"], are also infinite.

18. Typeface *designs* as such may not be copyrightable, but computer programs that generate typefaces are copyrightable and can be registered with the Copyright Office [4,5]. Screen displays generated by a program have long been copyrightable, suggesting that the displayed font may be entitled to protection, depending on its use.

substantially similar to another "a"? Perhaps we should use a topographical analysis: Does it matter that one is closed at the top and another is not? Do we consider the weight of the characters, thickness, textures, extrusion effects? Yes. And no [6]. Depending how you look at them, sound events, DNA sequences, and a series of computer-generated cartoon faces can be similar, and a series of faces can have more in common with a DNA sequence than another face series [7].

If "substantial similarity" does not depend on the quantitative *amount* of copying, how does one measure the qualitative *type* or *kind* of copying that will run afoul of the copyright laws?

In the creative arts, the primary playground of copyright law, there was a widespread, if tacit, understanding about the acceptable changes and uses of original materials. Only certain kinds of copying were done: generally, the context did not change at all (for example, a poem was copied in an anthology of poems), or, if the use *was* different, it did not differ by much. For example, a poem might be exploited as a song; a painting might be exploited in a photograph; pieces of original computer code might be found in a new program; or the use of one program's screen display might appear in a new work. These transformations were, traditionally, modest; the contexts were stable and mature. It followed that the law had little problem finding copyright violations—that is, that the "same" thing was found in the old and new work and that the original author's creativity had been exploited, or carried over, into the new work. In those cases, an examination of the new work *obviously* revealed the presence of the former, copyrighted work.

Things have changed.

Instead of traditional adaptations and derivations, we have in digital media the possibility of unlimited transformations of all the elements of work, together with the addition and subtraction of any property of a work. We can do it easily and cheaply, hence on a mass scale, and therefore the mutations cannot be controlled. A drawing of an orange skin can be the texture mapping for a new car; the lines of a poem can be extruded into a three dimensional object, then given properties such as mass, density, and resiliency, and then used by an architect as the building blocks of a house (using Autodesk's Cyberspace Developer Kit).

So, again, what *kind* of copying is prohibited?

Categories of Mutation: A Diversion

First, consider the types of changes or modification possible.

I think of the issue as one of data morphing. I use the term because it connotes a spectrum, a range of slight differences which at their extremes produce extremely different objects. In addition, the term "morphing" plainly suggests that *any* one item literally can be derived from another, and that, indeed, there may be a one-to-one correspondence between any two items: that one can, in a sense, be mapped to the other. If we needed reminding, morphing shows us that any two things can be thought as having similarities—but by the same token, *any* two dissimilar things can be morphed: Germs and judges, poodles and chocolate bars, peanuts and cosmic rays. That is the lesson of digital information.

I use the term "morphing" very loosely. I mean to refer to any shift or change in digital information that, in a given increment, has a definable relationship—usually one to one—with the prior incarnation of the datum. There are two general categories: (1) surface, and (2) internal or structural morphing.

Surface morphing often but not always involves structural morphing, and vice versa.

1. *Surface morphing.* This includes, for example, type extrusion and modeling programs,[19] which take Postscript or TrueType fonts and mutate them: by extrusion; adding 3D effects; rotation; texture mapping with a variety of other data; lighting from various angles; mapping a PICT image as a surface texture; or changing the shape, form, size, color, and other aspects of the item as presented to the senses. Similar tools have long been available to work with sound.
2. *Structural morphing.* This includes, for example, code type changes: JPEG, MJEP compression, fractal compression technology, changing bitmapped to vector graphics, and vice versa. In the sound domain, we would include transformation between various file formats designed for various sound cards. These shifts in data generally used by the processor of digital information, as opposed to the changes sensed by the user, can be termed structural morphing.

It is plain that the use of any kind of mutated data, whether the result of surface or structural morphing, can be the subject of—and barred by—copyright law, just as any type of audio-visual display or code (source code, object code, and micro code) can be copyrighted. And it is irrelevant that structural mutations take place without a visible change in the surface of the information, or that transformations take place in the

19. i.e., Pixar's Typestry and Strata Corporation's StrataType 3D.

appearance of information with only the most trivial changes—if any—in the data's structure.

It is equally obvious that the radical transformation of data will at some point result in creation of a new object. Every creation is, after all, the result, at some level, of the manipulation of preexisting items.

That fact—that creation is always a rearrangement of prior elements—is not remarkable. What *is* remarkable is what we now recognize as "prior elements." The analysis of artistic creation before computers focused on what I have called surface morphing and the relationships between surface elements; the advent of digital analysis not only created a new form of information, but, as importantly, a new way of understanding information. New *and* old works can be digitized; now *all* works have ascertainable structural elements—not only a new Kodak® CD-ROM photograph, but the old *Mona Lisa* as well—because we can transform analog data into a digital stream, to virtually any level of accuracy.

Previously we determined similarity by relying on rough abstractions from surface properties, such as character, plot, word structure, color, and shape. Obviously, that severely limited the kinds of changes that a work could undergo and still be treated as similar to a former incarnation. Tracing only surface morphing through a few changes, in other words, quickly leads to the conclusion that an entirely new work has been created. Take a painting, change the color and size, slice it in half and turn it upside down, and look at it through a glass lens: No 19th-century judge would easily conclude that the old and new work were "substantially similar."

But because we recognize *structural* morphing, we can trace what might be called the "same" data from appearance to appearance, from mutation to mutation, across various contexts, from font to texture to block to building. We can track the data like this either because (1) the digital data is in fact intact and has simply been fed through various programs and filters, or (2) we can plainly note a one-to-one or other algorithmic correspondence between the old and new data structures. And because this algorithmic structure really does exist, it lurks as a complete and powerful, but secret and silent, authority that *something* really is constant—is the same—as creations mutate and morph from one thing to another.

The implication of structural morphing of digital data, then, is this: There is no end to the types or severity of changes an object can incur and still be judged to be the same, or substantially similar, to its old embodiment. And the recognition of algorithmic correspondence can destroy copyright law, because when no possible kind of change is enough to defeat a copyright claim—when everything is a violation of the law—then nothing is a violation of law, and the idea of "copying" loses its force.

The Context of Meaning

Context; it depends on the context. Some changes, and not others, will *in context* appear to implicate the "essence" or "structure" of a work, and finding that essence or structure in the new work will prove infringement.[20] In other words, given that no mutation, no change from the original, can provide a defense to infringement, then one must abstract to some extent the work at issue to see if it can be found in substantial part in the new work. And what gets abstracted depends on the context. In poetry, perhaps rhythm and sound will play a key role; in a play, the dramatic structure and characters; in a painting—well, that may depend on the era—lines, color, subject matter; and in music, it may not so be much the notes themselves, or their timbre or color, but intervals between the notes that may take center stage as "substantial similarity" is debated.

This leads to a elemental problem. The single fundamental proposition of copyright law is that it protects actual, tangible expression; it does not protect abstract ideas.[21] But the test for infringement requires that expression be abstracted—that a judge precipitate the "essence" or "structure" to compare with another work. And that sounds very much like infringement can be based on an *idea* inherent in a work. At present, there is no way around this conundrum: to some extent, it *is* a work's idea—or better, one or more of the many ideas, or abstractions, of a work—which is protected by copyright law.

The problem is that everything is liable to differing levels of description, every object or work simultaneously can be thought of at different levels of abstraction. We look at a television display, and see colored dots on the screen, or Shirley Maclaine talking, or a piece of furniture. What is it actually? All of that [8].[22]

And the key issue of *which* aspect of a work, *which* idea, *which* abstraction, gets protection depends on the context and on whether the specific mutation is one that some judge thinks is central, in some way, to the work.

The advent of infinitely variable data morphing, where material can be fashioned into an entirely new class of objects, makes it difficult, if not impossible, to determine what the essence or structure of a work is, to know *which* description of a work will be used to determine whether that work has been illegally used by a new author. No one knows what the context is: architecture, art, photography, social interaction in a multiuser

---

20. *Feder v. Videotrip Corp.*, 697 F. Supp. 1165 (1988).

21. 17 U.S.C. § 102(b).

22. As every programmer knows, software can be treated at multiple levels of description: pseudocode, a collection of modules, the completed program, and so on.

dungeon, public and private dramatic works—the same data could be traced through each context, morphed as it passed through those domains, retaining its essence or not, depending on the extent to which one treats contexts as similar.

That is a summary of aspects of the current situation. For those involved in outright piracy, those who try to sell the work of others, the issues are not ambiguous: that is theft. But for others who bring original energies, time, and thought to fabrication of apparently new creations, copyright presents a dilemma. The problems are serious for the multimedia community. But the problems becomes intractable as we enter the virtual reality context.

COPYRIGHTING THE UNIVERSE: PROPERTY RIGHTS IN VIRTUAL REALITY

The notions of authorship and substantial similarity are central to copyright doctrine. Both are likely to disintegrate in the face of the technological and social revolutions conjured by virtual reality.[23]

By "virtual reality" I mean massively networked, broadband, interactive, immersive computer-mediated experience. The data, including audio-visual displays, feedback, and other sensory input are a function of (1) the operating system, (2) multiple layers of application software including those running on a host or hosts, and (3) programs running on remote platforms. Importantly, the kinesthetic interface will be a function of many users: both the users currently in the environment and probably prior users. A truly sophisticated VR environment will be extremely complex: it will be theoretically unpredictable; that is, chaotic, just as real complex systems exhibit rule-bound but chaotic behavior. Indeed, I suggest it is this unpredictability, these open-ended processes, that will make a crucial contribution to the felt perception that the experience qualify as a convincing virtual reality.[24]

In such an environment, we will find:

- The elimination of static, fixed objects capable of being owned, and the dissolution of identifiable authors making identifiable works;
- the merging of structure, surface, background, and fact.

---

23. To be sure, property rights, including intellectual property, in the VR context are likely to be overseen by developments in a variety of areas, including trademark, trade dress, patents, and the other areas I referred to in my introductory remarks.

24. See [9].

Let me briefly outline these two changes.

First, in multisensorial interactive VR, we are all authors and joint authors, broadcasters, writers, and consumers of data—importantly, we are all shapechangers; we are all responsible, to some extent, for the mutation of data—but because we do not completely control, and cannot predict, any change in the data, none of us is the identifiable author of any creation.[25]

Second, virtual worlds will be based on interconnected or identical operating systems, common communication protocols, object-oriented programming, common data structures in different locations, similar or duplicate interfaces and device drivers, and recognizable, common, standard sensory objects. These are the structure and surface of virtual reality, but both structure and surface will be part of the fact, the background of the world. Virtual reality, after all, is the context itself, the place where work and play get done—virtual reality is the multisensorial interface, the *environment.*

How much of this environment is uncopyrightable fact? Uncopyrightable idea? How much of the environment is uncopyrightable "scènes à faire?" At what level of description, or abstraction, will judges evaluate "substantial similarity"? Given our current view of VR as a fabricated immersive environment, I suggest that we will lose the contrast between object and ground, context and item. Judges will find it (virtually?) impossible to decide whether one virtual orange is similar to another—or to an enormous virtual red lobster, or an arm, or a virtual document—and judges are likely to be unable to meaningfully distinguish these virtual objects from the virtual environments of which they are inextricably connected. Because both surface and structural works are woven into a backdrop—itself a large, combinatory creation—copyright law will be unable to determine the "essence" or "structure" of an item, and the law will flounder.

But this may only be a first phase. If VR works and becomes an ordinary mode of human interaction, then the confusion of (1) environment with (2) objects in the environment may resolve itself. We may pass beyond an intense preoccupation with the environment; we may forget it, taking it for granted, just as we forget our current real environment. In short, over time, with a sufficient mass of people interacting, the virtual environment may become *stable.* In the midst of that virtual universe, small areas may be tamed, arranged, framed, ordered—that is, *created.* The essence, or structure, of those areas or pieces might be recognized by

25. "In cyberspace, everyone is the author, which means that no one is an author: the distinction upon which it rests, the author distinct from the reader, disappears. Exit author..." [10].

judges as the essence of protectible expression. And copyright law, itself morphed, may survive.

References

[1] "Pragmatism in Software Copyright: *Computer Associates v. Altai*," Note, 6 *Harvard Journal of Law & Technology* 83 (1992).

[2] Schulze, Herbert R., "Watch Out What You Wish For—You May Get Your Wish, or, Ownership Issues Continued: More on Applying the Work for Hire Doctrine to Computer Programmers," 8 *The Computer Law Association Bulletin* 12 (No. 2, 1993).

[3] Goodman, N., "Seven Strictures on Similarity," in *Problems and Projects*, Indianapolis: Bobbs-Merrill Company, 1972.

[4] 57 Federal Register 6201 (No. 35, Feb. 21, 1992).

[5] 53 Federal Register 21817 (No. 112, June 10, 1988).

[6] Hofstader, Douglas R., "Metafont, Metamathematic, and Metaphysics: Comments on Donald Knuth's Article 'The Concept of a Meta-Font'," reprinted in *Metamagical Themas: Questing for the Essence of Mind and Pattern*, 1985.

[7] Pickover, Clifford A., "Cartoon-Faces For Speech" *et seq., Computers, Patterns, Chaos and Beauty*, §§ 4.6–4.10, 1991.

[8] Hofstader, Douglas R., "Levels of Description, and Computer Systems," in *Gödel, Escher, Bach: An Eternal Golden Braid*, New York: Basic Books, 1978, Ch. 10.

[9] Morningstar, C., and F. Randall Farmer, "The Lessons of Lucasfilm's Habitat," *in Cyberspace: First Steps*, M. Benedikt (ed.), 1992, reprinted in *Virtual Reality 93 - Special Report (AI Expert)*, pp. 23, 25.

[10] Woolley, Benjamin, *Virtual Worlds: A Journey Into Hype and Hyperreality*, Cambridge, MA: Blackwell, 1992.

Copyright Issues on the Net: A Sampler

6

Many law firms have decided to maintain web pages, and so they should. These "pages" can be accessed by millions of Internet denizens, including many of our clients and would-be clients. Web pages can provide lawyers' resumes, copies of articles and newsletters, announcements of seminars, and pointers to other resources such as pending legislation, regulations, or other industry-specific sites. Web sites may allow for easy email communications with clients and others interested in our services. For many of our clients, having a web page provides a simple means to communicate with us.

THE MECHANISM OF A WEB PAGE

A web page is a compound document, accessible by anyone with an Internet connection. These pages are created similarly to ordinary documents—indeed, the current crop of word processors have the built-in capability to author web pages. Certain codes or tags embedded in the page invoke sound and visual images, which are played or displayed as the page is accessed by the viewer. More recently, soi-disant Webmeisters have begun embedding small programs or *applets* that download themselves to the remote client—that is, the viewer's computer—and then execute on that local machine. This allows dynamic, interactive viewing: the viewer can play games, see realtime stock quotes or weather updates, or a rotating three-dimensional model of a enzyme.

Most web pages will contain hotlinks; that is, areas on the screen that when activated by a mouse click will read a new page address and load the page at that address. The content then displayed to the user may reside on the same machine as the initial page or it may be on another machine, nearby or halfway around the world.

LAW AND DEVELOPING TECHNOLOGY: A DISJUNCTION

Life as a lawyer would not be complete without traps, minefields, and ambushes. For those interested in setting up a web page there are, of course, the inevitable legal snares.

In the world of Internet communications and other computer-mediated technology, it is now routine to note the friction between rapidly advancing technology and an increasingly ponderous legal system. That friction creates traps, making it difficult to predict the law's application not only to new factual contexts—the law does *that* all the time, reasoning by analogy—but to contexts where there is no obvious analogy, or where there are conflicting possible analogies.

In the Internet context, there is deep uncertainty on many issues. For those maintaining web sites, those uncertainties reach through copyright and trademarks to the legality of using encryption software to preserve client confidences. We know that there is no immunity on the Net for the transmission of obscene images, dissemination of trade secrets, or for the malicious distribution of viruses or other programs that interfere with one's use of data. But there is no consensus on what is obscene. There are no standards on the "reasonable precautions" one must take to preserve trade secret status. And there is precious little difference between (1) an ordinary computer virus, and (2) an applet that interferes with my intended use of my machine.

FOCUS ON COPYRIGHT

Copyright law on the Internet presents some of the more interesting traps for the unwary. Copyright is especially piquant because there is no *mens rea* requirement: There is no such thing as an "innocent" infringer. Errors are made at your own risk.

And everyone who sends email, maintains a web site, or participates in Usenet discussion groups, creates his own copyrightable material, and probably has appropriated materials owned by others. A recent book paints the picture [1]:

> This new thing, the Internet, consists of hundreds of thousands of linked computer terminals. Each individual computer can send and receive digital information—text and images—via the phone lines. And individual computer users can also link their own words and images with any of the millions of other computer users' words and images. Using this network, it is now

possible for people to access countless images. It's easy—perhaps too easy—to download any of these images to look at or to use. But it is ethical or legal to do so?

OWNERSHIP AND PERMISSIONS

The basic question for those who maintain web sites is ownership: Who owns the content the law firm (for example) has offered up for public viewing? Many firms hire specialized companies to program and create new materials, such as logos, graphics, and some text. Without a written agreement, none of this may belong to the hiring firm. That loss of control means similar content may appear on another firm's web site, and the original firm may have no rights to fire the programmer and still retain the content of their web site.

Other ownership issues are implicated when the firm (or its hired programmers) do not create, from scratch, the entire content of the site. Both the specialized companies that prepare web sites, and their customer lawyers who prepare content for the site, can be sorely tempted to use goodies found on the millions of computer sites that make up the Internet; or if not found there, to digitize the item by, for example, scanning it.

Scanned-in photographs and magazine articles—even those authored by employees of the firm—may belong to other entities. Those of us who work closely with certain industries—computer software, telecommunications, and so on—may be tempted to post industry-specific materials, perhaps secured from conferences or other Internet sites. That is usually a copyright infringement unless the owner has given permission or the use is "fair use."

Permission is usually needed, but permissions can be tricky. License agreements can be deceiving: even securing the *electronic rights* to a work may or may not cover one's use at a web site. License agreements may provide the right to use but not display, or to use in one context but not in another, or for only a period or time or for only a given geographical area (and, recall, the Internet is everywhere). Critically, permissions may not carry with them a right to *modify* the licensed materials. That makes the permission useless in the digital context where morphing, cutting and pasting, resizing, and other manipulations are *de rigueur.*

Permissions are often tricky, too, because it is difficult to know whom to ask. One of my clients wanted to produce a CD-ROM with a series of film clips: when advised that permissions from film houses, directors, actors, musicians, guilds, and all the rest might be necessary, the client gave up.

JOINT WORKS

Because web pages are what I term "compound documents," interesting issues pertaining to *joint works* may arise. Joint works have two or more contributors, each of whom has created copyrighted material and intends their contributions to be merged into a whole. Creating a web page with various copyrightable contributions of text, graphics, and perhaps sound files may trigger the joint work treatment. The problem is that without a written agreement, joint authors own an indivisible per capita share of the completed work, without regard to the amount of each person's contribution. So, for example, an artist who spends ten minutes designing the graphical look of a screen display and a programmer who spends ten months writing the code might each own 50% of the final product. This can lead to some unpleasantness.

To make things more complicated, the Copyright Act also recognizes *compilations* and *collective works:* with both, a copyright owner will own the larger encompassing work simultaneously with the various copyright owners for the constituent works. Figuring out who owns what can be impossible without a written agreement that attends carefully to the technology.

FAIR USE

One need not always ask for permission. Above, I used the copyrighted materials from Ms. Carter's book without asking. But my use is "fair." Use for quotation, comment, news reporting, and the like are all fair. Courts look to four factors: whether the use is for commercial purposes, whether the original is factual or creative (the latter gets more protection), the amount and substantiality of the portion used, and the effect on the market for the original of the copying. As with all tests that depend on balancing factors, we should be nervous of this one, since by definition the result cannot always be predicted. It is enough to note that the first factor is given great weight; by and large I suppose a web page devoted to law firm marketing is a commercial purpose. But not necessarily: This article and my Internet postings, I assure you, are for purpose of comment.

IMPLIED LICENSES

Even without asking, one might actually *have* permission, after all, by an *implied license.* The issue is nicely framed on the Internet: because the Net's denizens *know* that material posted at their sites will be downloaded by others—copied, in a word—there is surely is at least an implied license to make and use that copy. This is a sound argument, but it's difficult to know how far to extend it. What, in short, is the scope of the license? Look, don't touch? Are we to use the item once for personal use only? May one send the item on to others—as part of a larger work? And so on.

Some folks solve this problem by saying expressly in their material what can and cannot be done with it. But many of these express licenses are an incoherent mélange of inapplicable forms that do not help much.

Provenance is an equally difficult issue on the Net. Things are not what they seem. False attribution, impersonation, and spoofs thrive in cyberspace. "No one knows you're a dog" when you're on the Net. By the same token, it is often impossible to know whether one is viewing a work as it was originally made, with its original express license and original authorship attribution. While many of these problems can be solved with a variety of authenticating systems (beyond the scope of this chapter), it remains true that, like everything else on the Net, express and implied licensees cannot be taken at face value.

WEB LINK LIABILITY

Some have suggested that the creation of unauthorized links to another's site may infringe the copyright rights at the second site. The argument is that a copy is made—at least at the viewer's terminal—of the material at the second site, a copy that has been caused in effect by the creation of the link. I am dubious. A link is a pointer, an address, an instruction to the browser software to go find materials on some machine. And those materials, let it be recalled, were posted with the intent that the public view them, which is the precise result of the link. This is a good argument for an implied license. Other intellectual property issues are doubtless raised by links (such as trademark) but not, I think, copyright. But, I could be wrong. In early 1997 a Scottish newspaper was found to have infringed another paper's copyright when it presented the original newspaper's content via an Internet link. A variety of factors led to the court's ruling, including the apparent passing off of someone else's product as one's own. But it does not suggest that links are enough for copyright liability.

CONTRIBUTORY INFRINGEMENT AND VICARIOUS LIABILITY

For those who maintain web sites or otherwise provide Internet services and access, the most interesting and difficult issues have to do with one's liability for the evil acts of others. The Internet is an endlessly mutable, linked, distributed computing system; as such, every act of copyright and trademark infringement, trade secret theft, and libel involves the witting and unwitting contributions of many entities. Our online felons need the facilities of the companies at which they work, and of the service providers who are the "on ramps" to the Internet, among others.

Traffic through the Internet is very, very high. Large sites, such as those involved in home shopping, have many millions of hits and hundreds of thousands of emails per day. Sites that provide news feeds from the thousands of Internet "bulletin boards," known as the Usenet, may process millions of messages per day. These large sites cannot possibly police the content they dish up.

Smaller sites, such as those at firms, will not have that sort of traffic to handle, but that's both good and bad news. The good news is that smaller providers will have an easier time tracking activity at their site. The bad news is that, for just that reason, the smaller provider is more likely to be found to be contributorily or vicariously liable.

Law firms and others who maintain web sites may be contributory infringers—and liable for the malicious acts of others using their system—if they (1) know of the infringement, and (2) induce, cause, or contribute to the activity. While the "contribute" factor can be difficult to handle, it may be enough that the site owner simply retains some control and provides access to the Net. While this knowledge and control can be difficult to prove for a large commercial provider, it will be far easier to establish when the provider is a small company or firm.

Vicarious liability will be imposed where one has the right and ability to control the acts of the primary infringer and receives a direct financial benefit from the infringement. *Knowledge of the infringement is not an element.*

It will surprise no one that law firms may be vicariously responsible for infringements caused by their employees; the issue is whether that liability extends to the acts of nonemployees using the site. How much control must the firm exercise over those others? What is sufficient use of site to implicate the owner: storage of the illegal copy or use of the site's email system? Is it enough to use the site's links to enable the infringement? And what is a "direct financial benefit" under these circumstances?

Intriguing questions. God grant that you are not the test case.

One solution is to avoid being a conduit in the first place. That is, a firm could (1) ensure that all content posted on the site has been prepared (and screened) inhouse, (2) bar communications other than directly with the firm, and (3) forego links to other sites.

That policy would be effective, but it would eviscerate the benefits of the interactive distributed environment that is the Net. If we wish to use developing technology, as we should, we will live with uncertainties, making legal prophesy as best we can.

Reference

[1] Carter, Mary E., *Electronic Highway Robbery: An Artist's Guide to Copyrights in the Digital Era,* Berkeley: Peach Pit Press, 1996, p. xi.

Bibliography

Religious Technology Center v. Lerma (E.D.Va. Nov. 28, 1956).

Religious Technology Center v. Netcom On-Line Communications Services, Inc. (N.D.Cal. November 21, 1995); see *The Recorder* (August 6, 1996) for a report on the settlement of this case and the guidelines the parties agreed to on copyright on the Internet.

Campbell v. Acuff-Rose Music Inc., 114 S.Ct. 1164 (1994) (fair use).

Part IV
A Technical Interlude: Error in Computing Machinery

Other parts of this book have discussed the legal backdrop pertaining to computer technologies, and further parts will continue that review. The next chapter temporarily sets aside those legal issues and spends some time examining a few technical characteristics of computers and their programs. The point is that computers are not, despite some of their press, infallible. Of course once that is noted, few would disagree: We all recall apocryphal stories of computer errors and glitches, such as bugs in software causing the loss of phone service or crashed operating systems in personal computers.

But on reflection and inspection, we usually find that human error caused the problem and perhaps conclude that computers are, alone, generally error free. A good example was the failure of the late 1996 Russian effort to launch a Mars exploration mission, which failed 36.7 seconds into a mission that took 10 years and seven billion dollars (*many* more rubles) to prepare. A classic overflow error was caused when a guidance computer, fed spurious 64-bit data and expecting 16-bit data, tried to convert the former to the latter [1]. The guidance systems failed, and the rocket self-destructed. The program, of course, did just as its instructions dictated; human programmers made mistakes in assuming what the rocket's velocity would be, in including unnecessary lines of code, and in not anticipating the data format of the information being passed from one part of the computer system to another. They also made mistakes in trying to save money by not including rather customary error-protection routines.[1]

1. For more information on the role of a software anomaly in the destruction of the Ariane 5 rocket and payload, see http://www.cs.wits.ac.za/~bob/ariane5.htm

However, computers do not run "alone." A finding of human error is always double edged, because it indicts not simply the humans that manage—and are managed by—computers. It also condemns the systems themselves as arguably unmanageable. And that, in turn, may be laid at the feet of the human system designers, who themselves are working within the confines of hardware and operating systems; it also condemns the human managers who did not train their subordinates in the use of the computers, and so on, in turn, to an endless circle of constraints of human and machine responsibility for the errors in the daily functioning of computers. There is little gained by "blaming" humans or machines for specific faults.

A central theme of these essays is that there are *inherent* problems in using computers, problems that are a function of the *entire computing environment,* including the humans designers and users. These issues can be outlined by focusing on a specific machine architecture, and Chapter 7 examines types of parallel processing machines. The point is not that those machines in particular are riddled with errors; rather, the chapter simply seeks to provide a sense of the limits of the machine. The massively parallel processing (MPP) architecture described here may pass the way of all silicon; however, the point that machines are *inherently* uncertain will not lose its authority. That theme is carried forward in other chapters of this collection, too; in particular, the chapter on distributed artificial intelligence in Part V.

The risks of computing are inherent in the relationship between machines and human users, because the effects of computers and their programs cannot be fully predicted. That means that there is no such thing as *completely* training human users, that one cannot fully debug programs, and that no manufacturer can fully guarantee the performance of a machine. This also means that unexpected and sometimes damaging results will be encountered in a remarkably wide number of contexts—in practically all contexts in which we use the machines. And so the issue is reflected in many of the chapters in this collection outside this specific section on computing error, such as the articles on artificial intelligence and on criminal law. The ubiquitous nature of computer error is a powerful argument that such error is indeed intrinsic.[2]

---

2. A source cited in many of the essays in this collection is the Internet RISKS Forum, whose archives contain an extraordinary wealth of examples of computer errors and misapplications in almost every imaginable context. This is contact information provided by the Forum:

   The RISKS Forum is a MODERATED digest. Its Usenet equivalent is comp.risks.

   => SUBSCRIPTIONS: PLEASE read RISKS as a newsgroup (comp.risks or equivalent) if possible and convenient for you. Or use Bitnet LISTSERV. Alternatively, (via majordomo) DIRECT REQUESTS to

Reference

[1] Gleick, James, "Little Bug, Big Bang," *The New York Times Magazine* 38 (Dec. 1, 1996).

<request@csl.sri.com> with one-line, SUBSCRIBE (or UNSUBSCRIBE) [with net address if different from FROM:] or INFO [for unabridged version of RISKS information]
=> The INFO file (submissions, default disclaimers, archive sites, .mil/.uk subscribers, copyright policy, PRIVACY digests, etc.) is also obtainable from http://www.CSL.sri.com/risksinfo.html and ftp://www.CSL.sri.com/pub/risks.info
=> SUBMISSIONS: to risks@CSL.sri.com with meaningful SUBJECT: line.
=>ARCHIVES are available: ftp://ftp.sri.com/risks or ftp ftp.sri.com<CR>login anonymous<CR>[YourNetAddress]<CR>cd risks or http://catless.ncl.ac.uk/Risks/VL.IS.html [i.e., VoLume, ISsue].

Information Loss and Implicit Error in Complex Modeling Machines

7

INTRODUCTION

High-end networked computers, and those required to run interactive virtual reality environments, have inherent characteristics that may produce inaccurate results. "Inaccurate" in this context means a discrepancy in the comparison of computer or virtual worlds as compared to physical reality or direct mathematical calculation.

Complex modeling machines primarily are used to generate interactive computer-mediated or synthetic environments and for high-end scientific modeling. These applications require, in brief, (1) high-speed hardware architecture, generally using a form of parallel processing, and (2) a multilayered hierarchical architecture implemented in software. There are interesting consequences stemming from both the parallel processing and concomitant hierarchical processing paradigms.

The notion of parallel processing, which I expand to cover networked, distributed computing systems generally, generates these problems among others:

1. Strong reliance on the compiler and consequent compiler error;
2. The impossibility of reliable communications between processors, with problems including timing/computation communications overlap and fatal effects of processor crashes.

The notion of hierarchical processing also generates interesting problems which center on the fact that data is abstracted from one level of

The author thanks the staff and committee responsible for Networked Reality '94, the First International Workshop on Networked Reality in Telecommunications (May 1994, Tokyo, Japan) for the invitation to speak, which led to this paper.

processing to the next, and that, therefore, information is lost in such a way that the system is unaware of the loss.

These are consequences of classic computing machines. Whether serial or in parallel, these machines subject data to a sequence of algorithms. But there has also been much interest recently shown in cellular automata machines. I will also briefly discuss these systems in the context of a generalized overview of "error" as it pertains to complex modeling machines.

A GENERAL DESCRIPTION OF SYSTEMS

Complex modeling systems are currently used in a variety of applications of interest.

First, advanced interactive synthetic or *virtual reality* environments. Typically, a relatively expensive (from $150,000 to millions of dollars) machine is used in conjunction with a variety of peripheral devices (such as head-mounted display units, goggles, and gloves wired for input control) to allow the user to participate in a computer-mediated environment. Humans in other locations can participate as well. Applications include tours of virtual museums,[1] medicine,[2] military training[3] and conferences among, for example, engineers jointly viewing a virtual mock-up of a Boeing airliner. Relatively high computing power is required for these systems: they require rapid updating of the visual environment and the concomitant recalculation of complex objects rendered with lifelike textured, shadowed, and lighted surfaces, while at the same time accounting for physical laws such as gravity and air pressure, the characteristics of objects such as elasticity and refractive index, and of course the actions of the multiple humans in the environment.

Second, complex modeling systems are used as experiment test beds, as synthetic environments in which to review biological or physical hypotheses, and on the basis of such hypothesis to predict physical behavior. The benefits of computer modeling are plain: a variety of hypotheses can be explored, and none need be done in real time. For example, the environmental effect of predators on their prey can be explored in a complex, computer-generated world, moving through thousands of genera-

1. The Networked Virtual Art Museum uses virtual reality helmets, which are used to visit a virtual museum, simultaneously and interactively with other humans from around the globe [1].
2. See [2].
3. This includes the Close Combat Tactical Trainer and other joint combat operations with a variety of weapons platforms. See [3–5].

tions and millions of initial conditions—and all accomplished in an evening.[4] Applications include much of the work done under the auspices of the Los Alamos Center for Nonlinear Studies: high-energy physics (including nuclear weapons simulations), global climate modeling and geophysics, astrophysics, neural modeling, and a wide variety of other tasks.[5]

Increasingly, these machines are being used in commercial settings.[6] As I will conclude, the transfer of this technology from the defense, research, and academic settings to the commercial and consumer contexts[7] will pose interesting and difficult issues, both technical and legal.

For both these sorts of virtual environments, the most popular machines are those that provide the greatest computing power per dollar and that can simultaneously update a large amount of data. Those include massively parallel processing (MPP) computers with a distributed memory architecture—for example, the CM-5 (which is a network of SPARC processors), Cray T3D (with DEC alpha processors), and the Intel Paragon (with '860 processors) [11].[8]

As I suggested earlier, I will refer not only to the distributed processing found in classic MPP machines such as the CM-5, but also to distributed networked systems in which separated central processing units (CPUs) communicate. As the technology advances, I expect to see a blurring of lines among systems with (1) closely connected CPUs such as the CM-5 or the Maspar MP 1 and 2 machines, (2) computer *clusters*, encompassing physically neighboring machines [15], and (3) wide-area networks, made up of remote machines. High-speed communications, improvements in the classic input/output bottleneck, and the advantages of distributed databases all promise a closing of the performance gap among these three types of distributed systems.[9]

---

4. See [6].

5. See [7,8].

6. See [9]. (The traffic analysis required 33 hours of processing on a 256-processor Parsytec GCel-3.)

7. About 95% of MPP machines are devoted to scientific research of one kind or another; by 1997, the commercial sector will probably account for about 70% of MPP sales, with the market growing from $500 million to $5 billion [10].

8. The CM-5 (Connection Machine) actually uses a combination of distributed and shared momory architecture [12]. Late in 1993, Cray announced the CS6400 (4 to 64 SuperSPARC CPUs) [13], and the National Science Foundation has announced a study for a one trillion operations per second (TFLOPS) parallel machine [14]. Intel announced a TFLOPS machine in early 1997.

9. See [16], using both a Cray Y-MP computer and cluster of IBM RISC/6000-560 workstations. For the simulations here, the cluster performed from 3/2 to three times faster than on a single processor of the Cray. As chip processing speeds reach their physical limits, new efficiencies are likely to be found in parallel architecture. See also [17].

PROBLEMS OF TYPICAL PARALLEL SYSTEMS

At Inception: Compilers

A wide literature notes the difficulties of writing code to use parallel processors effectively.[10] Optimization in this context includes the actual assignment of processor tasks, and that is done in practice by compilers. Accordingly, we have quite literally the phenomenon of software writing software, "without the intervention of the user" [20]. There is generally no good way for the programmer to check the work of the compiler, because it is the compiler that generates the hardware synchronization operations [21].[11]

The issue here has been referred to as the *perturbation effect* [22], or the *synchronization effect*. Better put, these are variations of a central problem. The perturbation effect occurs when algorithms execute asynchronously: "any change in timing caused by the debugger can alter the execution order, perhaps masking a race condition or even creating one" [24].[12] In the normal execution of the program, synchronization of the various processing modules is also critical; this is discussed further in the next section. It is enough to note now that a program *as written* may be faulty precisely to the extent it seeks to harness large number of simultaneously processing CPUs [25]. This is an inherent condition of the debugging process: adding debugging code changes the timing of message passing among processors, and so masks basic timing issues [26]. Because complex software cannot be completely debugged, the result is potentially (but to an unknown extent) inherently unreliable software.

Problems in Concurrent Execution

Let us assume a shared memory architecture, whereby all processors address all areas of volatile memory. It turns out that serious problems arise when processors act independently on such dynamically changing shared information. When a communication channel breaks down between two processors, one of them will be uncertain as to the other's state; other processors may not be aware at all that a communications channel has been compromised [27].

10. See [18,19].

11. Debugging software has been announced that supposedly allows greater flexibility in handling some of these problems [22]. See also [23].

12. A "race" condition refers to multiple instructions to processors where the order of execution is critical.

While it may be possible for distributed computing systems to share memory—and indeed, there are MPP systems described as "shared memory" systems, as opposed to distributed memory systems—I suggest that at root *all* distributed computing systems, including tightly coupled CPU systems, must to some critical extent actually be shared distributed memory systems. Each CPU has at least its onboard memory, such as cache memory, and generally more than that as well. It is fair to assume, then, a distributed memory model of such systems.

In any event shared memory architecture does not allow parallel computing to fulfill its promise of harnessing enormous numbers of processors to undertake tasks involving extremely large amounts of data [28]. There may also be theoretical problems here.[13] Thus, it is most important to look at distributed architectures.

In distributed systems, substantial system resources must be allocated to the key function of communications among the processors.[14] The transmission of information is critical, and its failure is of course a failure of the entire system.

As it happens, no reliable communications protocol can tolerate crashes of the processors on which the protocol runs. This is because information must be able to survive such a crash, and such information cannot survive a crash in volatile memory [29]. Indeed, no protocol reliably can provide either delivery or notification of nondelivery for all messages, including those sent before a crash.

Thus, nonvolatile memory must be used to store information. While it has been suggested that the hardware clock could act as such a storage medium [29], others have noted that clock synchronization is imprecise and subject to vagaries of operating systems and onboard caches, among other things [30]. Disk storage is the next candidate for information storage. However, while that is probably reliable, it is probably impossible to use anything like that medium to store information generally found in

13. Even with a shared memory system, allowing every processor to address every region of system memory will destroy the integrity of a system state *unless* there is some other portion of the system that blocks a given write to memory when (for example) another task needs to be executed first. That "other portion" may need independent and volatile memory.

14. A Connection Machine has 65,536 processors and associated memory, and thus contains the equivalent of a telephone switching system handling 65,000 subscribers—each of whom places thousands of calls per second. The Maspar MP-1 includes 16,384 processors.

very high-speed CPU registers, buffers, and cache memory.[15] By default, volatile memory is used to coordinate processing in distributed memory systems.

Thus, we may lose information during the operation of an MPP system without the assurance of communications reliability among processors.

GENERAL PROBLEMS OF HIERARCHICAL SYSTEMS

abstract,...L. abstractus, drawn away, withdrawn...[thus] take away.

—*The Oxford Dictionary of English Etymology*

The hierarchical form is generally implemented as (1) layered, interacting applications, and (2) within a given application, a multitude of logical levels separated and defined by branching instructions, subroutines, modules, and the like. In this context, the distinction between (1) multiple, layered applications and (2) multiple logical levels within an application is not significant. In each case, we note the iterated mutation of data. In that mutation, information is of necessity lost.

At Inception: Creating the Program

The data structures and algorithms of every program reveal assumptions about the real world. No model can be completely correct, and that is recognized as a "fundamental difficulty of systems design" [31]. Even subtle models are more crude than reality, which is infinitely complex and which can be analyzed at an arbitrary depth of detail.[16] Programmers hope that the essential behavior has been modeled, but that will not always be correct, and there may be no way by which to determine whether one is correct or not.

---

15. That, at least, is the present situation. Advances in hologramic storage and other light-based computing may provide very-high-speed nonvolatile memory. As the introduction to Clifford Pickover's *Visions of the Future: Art, Technology, and Computing in the 21st Century* notes, "In 1993, researchers at the University of Colorado built the first fully optical computer that stored, shuttled, and processed its data entirely in the form of light."

16. See [32]. "The fundamental limit on the accuracy of simulations is the inherent complexity of reality itself" [33]. The use of constraint programming to deal with this issue is discussed in the section entitled "Turning to Cellular Gas Automata" later in the chapter.

The physical world is of a continuous fabric, both across its features and in depth. As that world is digitized, small but critical details can be lost. For example, the physical world has its constraints on the use of the gas peddle and fuel throttle, but when those are replaced by digital simulacrae, artificial constraints must be devised which may or may not cover the full complex (and probably unpredictable) range of behavior of the interplay between a real throttle and brake [34].[17]

If the "system" is thought to include the human participants, as it probably should, and as it certainly is in virtual or synthetic reality systems, then the overall behavior of the system becomes unpredictable; so much so that "error" must be expected. Indeed, "error" is the means by which the system stabilizes through feedback from the human, and "error" is essential then to the proper operation of the system [36,37].

The Operation of a Hierarchical System

There is nothing new about noting the abstraction, and thus information loss, that goes into the creation of a program. But little attention has been paid to the progressive abstractions that occur within the computing system as it operates. In its very operation, a hierarchical system constantly abstracts and thus loses data.

First, we have the rounding problem. Infinitely repeating fractions are found in the usual base ten and computer *lingua franca* base two representations (i.e., 0.1 is 0.00011001100...). If this is rounded off, plain error results.[18] Thus, at the most basic level of computation, there is a serious risk of information loss whenever irrational numbers are involved. How serious can this be? Rational numbers are islands in a sea of irrational numbers;[19] in the modeling of complex phenomena, where the slightest distinctions in initial conditions can spell vast differences in output, rounding off irrational numbers may make all the difference in the world.

Second, we note the structural hierarchies in a program. These hierarchies are traversed from input to device driver, past networking programs to a series of modules within an application that communicate with the hardware through a variety of kernels, operating systems, and so on. At each level, information is passed through some form of an inter-

17. There is a good discussion of this issue in the aircraft control context found in [35].

18. Apparently, Borland's C++ version 4.0 does just that [38].

19. As George Cantor showed, there are very many more, an order of infinity more, irrational numbers than rational numbers; there is an uncountable infinity of irrationals, as opposed to the merely(!) countable infinity of rational numbers.

face, and, to complicate the picture, generally the information flow is bidirectional.

This hierarchical operation is not designed to (and does not) generate a one-to-one correspondence between the data input and output across an interface. Data is not "transformed" as in the senses of an affine transformation or other transformations which preserve a one-to-one mapping relationship. Data is not maintained as the passing of a variable from one subroutine to another. Instead, the systems take enormous quantities of data and generate an *abstraction* of it. That, indeed, is the point, the core intent, in the development of these systems. The essence of a hierarchical processing system is to exercise some form of intelligence over the raw input, produce conclusions based on the current algorithms, send on the conclusions—which are then treated as the raw data for the next level of processing—and so on iteratively.[20]

This phenomenon exists from the lowest to the highest logic levels: from the use of input as a counting parameter to generate a binary conclusions (e.g., more than three unsuccessful attempts to access a database warns of an attempted illegal intrusion) through fuzzy logic (e.g., a variety of input suggesting a "dangerous" overload on an electrical system) to large-scale synthetic environments with input from humans, abstract physical laws, and measured data from portions of the real world. In all of these cases, a sort of pattern recognition takes place, and all pattern recognition implicates the truncation, or loss, of data.

This picture of computation is substantially different from the traditional "input > algorithm > output" model—a cartoon of the classic Turing machine, in which computation is the mechanical application of mathematical transformations. The process is not "mechanical" in this way; the results are best thought of not as mathematical results, but rather as hypotheses about the world or about some possible world.

This aspect of large computing machines is well illustrated as we consider reactive systems (that is, computers which react to signals, conditions, and other input in complex ways). Examples include chemical and nuclear control systems, avionics, some robots, sophisticated virtual reality systems, and even common programs such as word processors [40]. These systems have large numbers of possible states, and accordingly it can be extremely difficult to create programs that always "work" or produce reasonable (and safe) results. If, unlike the classic Turing ma-

20. Arguably, the same process takes place in the brain as it ingests and integrates an enormous range of input to reach conclusions and to act [39].

chine, these may be infinite-state machines,[21] then the problems are concomitantly intractable.

But complex reactive systems have a further problem that illustrates the degree to which their processing in fact transforms and eviscerates information. David Bella has mapped out state space of reactive systems, and notes that those with high connectivity and reactivity[22] very rapidly generate situations in which relatively minor events are amplified throughout the network and lead to wholly unpredictable, and thus possibly catastrophic, results [41].[23] Again, the output of the reactive network, which can include distributed systems generally, is not commensurate with the input: information is lost, and new information is created. This can be thought of as a concomitant of the parallel or of the hierarchical processing models: here, they amount to the same thing.

MPP machines are, after all, designed to manage very large amounts, veritable blizzards, of data: to be useful to the human operator, the output of necessity refines the data—which is a pleasant way of saying that data is eliminated. Some current thinking on the uses of enormous amounts of data and processing capabilities celebrates this abstraction of information.[24] That confidently predicts the existence of massively parallel and extensively recursive software ensembles—"mirror worlds"—that will be able to process and model data about our social, political, economic, and physical world and present it in a computer-mediated abstraction. This vision may be theoretically feasible. And there may indeed may be no better way to handle massive amounts of information. But, based on the problems outlined earlier, there is no assurance that these mirror worlds would mirror the worlds we know (see note 21). Accuracy is not assured; indeed, error is virtually guaranteed.

---

21. Traditional computers are thought of as finite-state machines. Given the general equivalence of universal computing machines, one would assume that complex computers may have a large but still finite number of possible states. But recalling that state space is a function of the properties one chooses to map, I suspect there are infinite state machines—but even if not, (1) highly complex machines have more states than it is possible to attain or describe, and (2) it may not possible in advance—algorithmically, as it were—to know whether or not a complex machine has an infinite number of states. I note a corollary here, which is that two infinite-state machines cannot be mapped to each other. Practically, this noncongruence may be as valid for any two complex systems, because we might never know nor have the ability to know whether the two systems could be mapped to each other.

22. Reactivity measures the degree to which the activity of one node depends on that of neighboring nodes.

23. See also [42], which notes that the addition of noise to multistable systems can amplify small period signals inducing transitions between the two states.

24. See [43]. "Forget about the details," Professor Gelernter suggests on p. 125.

TURNING TO CELLULAR GAS AUTOMATA

Once endowed only with strict mechanical, calculator-like functionality, computers now are used to simulate and manage complex portions of the real world. That involves making and implementing hypotheses about the world. That generally means hierarchical levels of processing, which entails information loss, and consequently the unpredictable risk of error.

Perhaps a result of an implicit understanding of these shortcomings, much attention recently has turned to a different approach, known as *cellular automata*, or lattice gas or Turing gas machines. These systems involve enormous numbers of objects changing their states through interaction with neighbors, according to a few general rules. These systems are special creatures of MPP systems in that they are most efficiently implemented on parallel systems, and they apparently simulate what are in the real world systems operating in parallel, such as turbulent flows, heat flow, and other fluid dynamics.[25]

The difference between these systems and those discussed above is this: in lattice gas systems, "computation proceeds by the propagation of constraints rather than by the execution of an algorithm" [48]. Extremely complex behavior emerges from simple systems evolving concurrently on a very large scale; some have termed it "self-organization."[26] This sets the lattice gas systems apart from classic algorithmic systems, because the lattice gas computation loses no information in the sense described earlier: there is no loss as a result of representing numbers in a machine [50], and the input and output at any iteration are qualitatively identical. Lattice gas machines are strongly parallel, but not hierarchical.

Programming these new systems, however, introduces a new source of "error." One seeks to "model" the phenomena of interest by trying out various series of general rules, viewing the evolution of objects that results, and then modifying the constraints until the output appears correct. The "successful" constraints are then announced as part the correct model of the phenomenon studied. But how is success measured? What does it mean to announce that "lattice gas/cellular automata methods... show the qualitative behavior of complex systems with astonishing *fidelity...*" (emphasis mine) [51]? The answer is: this may not mean much.

25. A simple example is Conway's Game of Life, consisting of a two-dimensional table of points, each of which lights on or off in one state depending on the prior state of its neighbors. See generally [44–46]. While cellular automata/lattice gas systems run best on MPP systems, they also can run on traditional serial or von Neumann machines, as all of these different systems (and others) are, finally, equivalent. See [47].

26. See, for example, [49].

Perhaps the best that can hoped for here is the conclusion that *functionally* the lattice gas machine has accurately modeled the physical phenomena; no detailed information about a real complex system will be generated [52]. The *patterns,* we say, are the same or similar or analogous. This can be thought of as the inverse of the information loss vector described earlier: abstraction and loss not as a consequence of processing, but as a presumption, a self-fulfilling prophecy, of the processing. There is no pretense here: the results from a lattice gas computation are unabashedly and inherently abstractions that simply *may* have explanatory power. It is up the human operator, finally, to decide whether two patterns are sufficiently alike to announce a "solution" to a problem.

True, some of these solutions will have predictive power, as when aircraft wings and economies behave just as the constraints suggest. But unresolved is the extent to which a past correlation between large complex systems (i.e., the physical phenomena and the lattice gas) is a reliable indicator of future correlation. For reasons outlined earlier,[27] I suggest this type of inference remains a difficult and unresolved problem.

SIMULATIONS AND MODELS

This analysis suggests making the distinction between the terms *simulation* and *model.* A simulation *looks like* the phenomena imitated: its actions and surface appearance are, for some limited purpose, indistinguishable from the phenomenon. Flight simulators fit this category: "accuracy" is provided only to a certain degree; no one would confuse the simulation with the real thing. By contrast, a model is hooked into the real world, or purports to imitate real-world processes, down to any potentially meaningful level of detail. Electronic fly-by-wire flight systems in real aircraft such as the Airbus and the U.S. Space Shuttle, and real-world nuclear power plant control stations are within this category, as are models of structural stress and turbulence, where the solutions have predictive power. For many applications it is uncertain, and will remain so, whether models or simulations have been created.[28] The validity of a true model as defined here depends on proof that the phenomenon can *be completely encoded.* That proof probably depends on (among other

27. In a different context, others have also expressed serious doubt that meaningful content can be given to "similarity" sufficient to validate inductive reasoning on software reliability [53].

28. Note, for example, attempts to model prebiotic phases of evolution, the birth of the universe, and so on. While some putative models of the origins of the universe predict the structure of galaxies and supergalactic objects, other sets of constraints may do so as well.

things) the feasibility of defining the phenomena as a closed system. But many complex systems are not closed systems.

To rephrase, then, a conclusion reached earlier: we do not know when the computation has accurately modeled the phenomena and when it has simply made a good simulation of it.

CONCLUSION

There is justification for great enthusiasm over the enormous power of current machines and the capabilities predicted for the future. There is special excitement at the power of massively parallel processing machines: that architecture, including distributed processing environments, is and will continue to be the standard tool for many of the "hard" science problems. And it is very difficult to imagine synthetic or virtual reality environments operating on anything else. Networked processing and related networked distributed databases hold center stage in these important applications.

But there are profound—and I have suggested *inherent*—differences between physical reality and the synthetic worlds within machines. The very act of creating programs and inputting data abstracts and mutates the information. The processing of data, alone, mutates it. Not knowing when a program is a just very good simulation of physical phenomenon, as opposed to a true model of it, is important. It means we cannot be certain when the putative model will give way, when it will lose its predictive power, when it will cease to be reliable.

These issues will become increasingly problematic as parallel processing and complex modeling machines move from academia, research, and defense to the consumer milieu. As distributed systems are assigned the tasks of running real-world applications—as models are implemented—system failures will generate a new level of dispute. That new level is, in a phrase, the *legal* context.

Testing, failure, and rough results on the rapidly developing edge of technology are routine in the academic and defense contexts. By contrast, prudence and safety are often paramount in the commercial context.[29] Already large networked machines have caused failures with adverse economic consequences.[30] While uncertainties in the absolute performance

---

29. This conflict is described in more detail in [54].

30. The Usenet conference comp.risks collects and discusses a wide variety of computer-related problems, especially errors in real-world applications. Some have argued that, due to the inherent risks of complex software, the use of computers should be restricted whenever safety is of primary concern [25,55]. Contrast David Gelernter, who argues in

of a putative model is acceptable in the research context, the commercial world is not as likely to understand and forgive data loss—and loss of life and property—caused by failures of complex machines. These real-world, networked applications are likely to be judged by a standard unfamiliar to the researcher; i.e., whether the program works, and whether fair notice was provided to lay users and others on the specific risks involved. These may be difficult standards for the developers and purveyors of inherently and unpredictably error-prone systems.

References

[1] "Double Your Fun," *NewMedia* 119 (Sept. 1993).

[2] Satava, R., "Telepresence Surgery: Medical Implications for Virtual Reality," 2 *Virtual Reality Report,* (Nov. 1992).

[3] Moshell, Michael, et al., "Virtual Environments for Military Training: SIMNET, *Ender's Game,* and Beyond," 1 *Virtual Reality World* v (Summer 1993).

[4] "Industry and Military Show Capability," 13 *CyberEdge Journal* 14 (Jan./Feb. 1993).

[5] Gembicki, M., and D. Rousseau, "Naval Applications of Virtual Reality," AI Expert, *Virtual Reality '93* 67.

[6] Holland, John H., *Adaptation in Natural and Artificial Systems,* Cambridge, MA: MIT Press, 1992.

[7] Hillis, W. Daniel, "What Is Massively Parallel Computing and Why Is It Important?" in *A New Era in Computation,* N. Metropolis and Gian-Carlo Rota (eds.), Cambridge, MA: MIT Press, 1993, p. 1.

[8] Nagel, Kai, et al., "Deterministic Models for Traffic Jams," preprint, University of Cologne Appl. Math, August 6, 1993.

[9] "AMEX Extending Credit for Parallel Processing Plans," *Computerworld* 1, 14 (Dec. 20, 1993).

[10] *Computerworld* 14 (Dec. 6, 1993).

[11] Sharp, Oliver, "Compilers for Parallel CPUs," *Byte* 97 (Feb. 1993).

[12] Hillis, W. Daniel, et al., "The CM-5 Connection Machine: A Scalable Supercomputer," 36 *Communications of the ACM* 31 (Nov. 1993).

[13] *Computerworld* 10 (Nov. 1, 1993).

[14] *Computerworld* 77 (Nov. 1, 1993).

[15] "Computer Clusters May Be the Answer," *Computerworld* 13 (Dec. 6, 1993).

[16] Bonca, J., et al., "Quantum Simulations of the Degenerate Single-Impurity Anderson Model," Los Alamos National Laboratory, *CNLS Newsletter* (No. 96, Dec. 1993).

[17] *Call for papers and tentative description of the Journal of Computer & Software Engineering, Parallel Algorithms & Architecture,* reprinted, 25 *SIGACT News* (ACM) 61 (March 1994).

[18] Thomborson, Clark D., "Does Your Workstation Belong on a Vector Supercomputer?" 36 *Communications of The ACM* 41 (Nov. 1993).

[19] Harel, David, *The Science of Computing,* Reading, MA: Addison-Wesley, 1989, p. 233 *et seq.*

---

his *Mirror Worlds* [43], that highly complex real-time systems and vast amounts of data can *only* be controlled by hierarchical software layered into increasingly abstracted levels.

[20] Deng, Yuefan, et al., "Perspectives on Parallel Computing," in *A New Era in Computation,* N. Metropolis and Gian-Carlo Rota (eds.), Cambridge, MA: MIT Press, 1993, pp. 31, 46.

[21] Hillis, W. Daniel, et al., "The CM-5 Connection Machine: A Scalable Supercomputer," 36 *Proceedings of the ACM* 31, 38 (Nov. 1993).

[22] Sorel, Pierre, et al., "A Dynamic Debugger for Asynchronous Distributed Algorithms," *IEEE Software* 69 (Jan. 1994).

[23] McKenney, Paul E., "Selecting Locking Primitives for Parallel Programming," 39 *Communications of the ACM* 75 (Oct. 1996).

[24] McKenney, Paul E., "Selecting Locking Primitives for Parallel Programming," 39 *Communications of the ACM* 69 (Oct. 1996).

[25] Wiener, Lauren, *Digital Woes,* Reading, MA: Addison-Wesley, 1993, p. 136.

[26] Sharp, Oliver, "Compilers for Parallel CPUs," *Byte* 99 (Feb. 1993).

[27] Birman, Kenneth P., "The Process Group Approach to Reliable Distributed Computing," 36 *Communications of the ACM* 37, 40 (Dec. 1993).

[28] Deng, Yuefan, et al., "Perspectives On Parallel Computing," *A New Era in Computation,* N. Metropolis and Gian-Carlo Rota (eds.), Cambridge, MA: MIT Press, 1993, p. 49.

[29] Fekete, Alan, et al., "The Impossibility of Implementing Reliable Communications in the Face of Crashes," 40 *Journal of the Association for Computing Machinery* 1087 (Nov. 1993).

[30] Birman, Kenneth P., "The Process Group Approach to Reliable Distributed Computing," 36 *Communications of the ACM* 41 (Dec. 1993).

[31] Wiener, Lauren, *Digital Woes,* Reading, MA: Addison-Wesley, 1993, p. 46.

[32] Robinett, Warren, "Synthetic Experience: A Proposed Taxonomy," 1 *Presence* 229 (Spring 1992).

[33] Robinett, Warren, "Synthetic Experience: A Proposed Taxonomy," 1 *Presence* 245 (Spring 1992).

[34] Wiener, Lauren, *Digital Woes,* Reading, MA: Addison-Wesley, 1993, p. 154.

[35] Dorsett, Robert D., "Risks In Aviation," 37 *Communications of the ACM* 154 (Jan. 1994).

[36] Sheridan, Thomas, *Telerobotics, Automation, and Human Supervisory Control,* Cambridge, MA: MIT Press, 1992.

[37] Mindell, David, "Telerobotics, Automation, and Human Supervisory Control," Review, 12 *IEEE Technology and Society Magazine* 7 (Fall 1993).

[38] Coffee, Peter, "Close Enough Isn't Good Enough in Computer Math," *PC Week* 60 (Feb. 14).

[39] Minsky, Marvin *The Society of Mind,* New York: Simon & Schuster, 1985, p. 95.

[40] Harel, David, *The Science of Computing,* Reading, MA: Addison-Wesley, 1989, p. 265 *et seq.*

[41] Bella, David A., "Rethinking the Unthinkable," *IEEE Technology & Society Magazine* 9 (Fall 1993).

[42] Carroll, T. L., et al., "Stochastic Resonance and Crises," 70 *Physical Review Letters* 576 (Feb. 1, 1993).

[43] Gelernter, David, *Mirror Worlds,* New York: Oxford University Press, 1992.

[44] Eigen, Manfred, *Laws of the Game,* New York: Harper Colophon, 1981, p. 178.

[45] Poundstone, William, *The Recursive Universe,* Chicago: Contemporary Books, 1985.

[46] Dewdney, A. K., *The Armchair Universe,* New York: W. H. Freeman & Co., 1988, p. 133.

[47] Knudsen, Carsten, et al., "Information Dynamics of Self-Programmable Matter," LA-UR 91-0896 preprint, Los Alamos National Laboratory, August 6, 1990.

[48] Hasslacher, Brosl, "Parallel Billiards and Monster Systems," in *A New Era in Computation,* N. Metropolis and Gian-Carlo Rota (eds.), Cambridge, MA: MIT Press, 1993, pp. 53, 54 .
[49] Fontana, Walter, "Functional Self-Organization in Complex Systems," No. LA-UR-9-3924, Los Alamos National Laboratory, 1990.
[50] Hasslacher, Brosl, "Parallel Billiards and Monster Systems," in *A New Era in Computation,* N. Metropolis and Gian-Carlo Rota (eds.), Cambridge, MA: MIT Press, 1993, p. 61.
[51] Hasslacher, Brosl, "Parallel Billiards and Monster Systems," in *A New Era in Computation,* N. Metropolis and Gian-Carlo Rota (eds.), Cambridge, MA: MIT Press, 1993, p. 63
[52] Fontana, Walter, "Functional Self-Organization in Complex Systems," No. LA-UR-9-3924, Los Alamos National Laboratory, 1990, p. 11.
[53] Littlewood, Bev, et al., "Validation of Ultrahigh Dependability for Software-Based Systems," 36 *Communications of the ACM* 69, 73–75 (Nov. 1993).
[54] Karnow, Curtis, "The Uneasy Treaty of Technology and Law," in this volume.
[55] Littlewood, Bev, et al., "The Risks of Software," 267 *Scientific American* 62 (Nov. 1992).

Part V
Law at the Man/Machine Interface

And the LORD appeared unto him in the plains of Mamre:
and he sat in the tent door in the heat of the day;
And he lift up his eyes and looked, and, lo, three men stood by him:

—Genesis 18:1-2

William Gibson's 1984 novel *Neuromancer* introduces us to the prime mover of cyberspace, an agency of change, of motivation, master of its environment, an algorithm made flesh: its name is *Wintermute*. Wintermute is an artificial intelligence (AI), whose two-word name reminds us of the project codes generated by the computer in the Pentagon basement that brought us Tacit Rainbow, Senior Crown, and other secret project names.

Above and beyond the law, Wintermute kills, transfers money illegally, and abets assaults, mayhem, robbery, and trespass. Wintermute is too big, too complex, too encompassing to be threatened by law enforcement personnel, the Turing Registry. The only possible solution is to pull the plug, but that would kill everything else too: Gibson sets the final confrontation on an orbiting space colony, a entirely networked environment that controls, among other things, life support systems. If Wintermute goes, we go.

Wintermute is everything—everything in the universe of emerging computing power. It is the unknown, that which is beyond our control. It is natural and entirely human to *name* it, and so we do, much as Adam and Eve named the animals about them, providing the apparent means of definition and thus of control. And as powerful as the act of naming is its concomitant: making the unknown into our own image, making it appear something like a human. And so Wintermute and its dopplegangers, the other AIs in Gibson's book, manifest as personalities, people, in shadowy bodies, images on television sets and in cerebral electrode-induced hallucinations. They do not exist, but these young boys and old men have eyes, broken yellowed teeth, waving black hair, bittersweet laughs, and like us they sometimes ask for death. They wave good-bye and pick their teeth with wooden toothpicks. We can deal with that.

Our computer networked environments are replete with these fictive entities. Microsoft has "named" an operating system interface "Bob," to make us more comfortable with the software. At this writing, the latest craze among teenage boys in Japan is a pop star who appears on television, gives interviews, makes recordings, and responds to fan mail—but does not exist: "she" is a multimedia character, a figment of many imaginations. In those electronically mediated universes known as multiuser dungeons (MUDs), where real humans chat and meet, some of the characters are just software: they help to populate the world and give the alien machine worlds a recognizable texture.

Too, the terms "agent" and "wizard," with their deliberate connotation of personality, are used by a wide variety of software products to denote various help facilities; again, these act as intermediaries between the human user and the machine. The reader may also be familiar with the new crop of putatively intelligent agents, described in more detail in the chapters that follow, among other tasks designed to navigate the mazes of the Internet and retrieve hidden nuggets of information. A few recent patents expressly refer to the notion of computer read-only memory (ROM) personality constructs, using them to guide data searches and make the man/machine interface easier for the human users to navigate.[1] And while some readers may not have met Wintermute, many know HAL, the foremost computer personality of the popular imagination, a machine intelligence with indeed more personality than the humans who populated the movie and book *2001: A Space Odyssey* [3,4].

We use the image of a fictive person to stand in our stead, too. From time to time I receive email not from Phil Z., who is a friend and client, but from his email 'bot, who informs me that Phil is out of town and will read my mail later. We can adopt a variety of electronic disguises in our email, contributions to electronic bulletin boards, chat groups, and as we play interactive games. We know who we are, of course, but no one else does because the manifestations of these fictions cannot be distinguished from the authentic appearances or from the guises of software.

These digital personas are our Virgils. They shelter us from otherwise inhospitable environs. We mask the operations of computers with the veneer of human attribute. We *mean* to have a confusion; we intend to

---

1. See, for example, [1] and also [2], which ranks search retrieval results in an order determined by a "bias" of a specific encoded personality or character.

fabricate some recognizable entity that will guide us; we mean to pretend that the alien digital context is in fact familiar. From Dr. Frankenstein's creature to HAL 9000, we have desperately tried to put a human face on the machines and technology that seem to be invading our hemispheres.

♦ ♦ ♦

There is another important strand in the development of the human/machine interface, in addition to the development of digital personas. This is the increasingly sophisticated and convincing work of virtual reality. High-end computers are used in a wide variety of applications involving massive amounts of data, which is difficult if not impossible to absorb and manipulate without virtual reality technologies. These technologies handle and display enormous amounts of data in real time, attempting to create fully immersive computer-generated simulated environments that display the interior of aircraft, human bodies for simulated surgery, and complex molecules viewable from any perspective. Large amounts of data can be viewed as three-dimensional graphs over which one can "fly" and inspect from a variety of perspectives. Many of us are familiar with virtual reality entertainment centers, which put us "in" a computer-generated environment to produce a thrilling ride. The technology is so good that first-rate conservative airlines such as Swiss Air train their pilots in virtual cockpits—pilots whose first real flight in a passenger airplane is with real passengers.

To generate the immersive experience, the traditional accouterments of computer use are disposed of—keyboards, the usual monitors, and so on are all discarded. The user dons headgear, and perhaps gloves and other interface elements, and then navigates with bodily gestures, as one does in the physical world. Newer technologies dispense with that hardware as well, scanning images directly onto the retina of the eye and directly translating brain waves into computer commands. The point is to dissolve the man/machine interface; to eliminate computer hardware as an intermediary between the user and the products of computation. The goal is, as earlier chapters have termed it, the zero user interface (ZUI). The power of the machine is harnessed, as never before, with the purpose of making it vanish.

♦ ♦ ♦

The two technologies of electronic personalities and virtual reality meet. There is a good deal of work being done on virtual humans designed to populate virtual reality: These avatars test machines, walk through simulated mock-ups of Boeing airliners, act as patients in simulated surgeries, and test military doctrine in simulated battles. We have developed [5]

> human-shaped, intelligent agents that interact with each other and with real humans in virtual environments. The goal is to populate simulations with many dynamic, interactive, autonomous avatars.

Many developments in artificial intelligence research can be seen as aiding and abetting the search for ZUI. Work towards speech recognition, visual discrimination, programming in common sense and reasonable heuristics, and so on, will make it simpler, more "natural," for humans to interact with computers—for what could be easier than interacting just as we do with the humans we meet every day? There is a powerful desire for the ZUI, to have "computers improve to the point where talking to your computer is as easy as talking to another human being" [6]. The interest is not merely academic, for computers are no longer the province of scientists and technicians. Artists, linguists, and shirt cleaners all use computers, and they have no necessary love of the machine. Many do not want to use a "machine" at all, and to that end both the popular imagination and some the best of our research work is devoted to the "intelligent interface" [7]. That usually means ZUI, in which the electronic world is incarnated in the guise of the familiar physical world.

Of course, not all human/machine interfaces are modeled on electronic personalities or avatars, and more centrally there are a variety of other models that we could use. At least in the abstract, it seems as if we could visualize computers as pets, wild animals, clockworks, or utilitarian artifacts such as bowls and baskets. We employ many other machines without serious efforts to anthropomorphize them: we use cars, stereo equipment, automated factories, and toasters without the felt need to imagine them in our human image. But other chapters in this book suggest that computers, at least when used as general-purpose computing machines, are inherently different from these other technologies and call out our need to equate computers and minds. For now it is enough to remark that we are strongly tempted to that equation, and that important developing technologies, including virtual reality, intelligent agents, and AI research generally, will further it.

But once we have raised these human constructs, what then? How far do we take the conceit? Quite a way. Outside of severely limited areas, computers and their instantiations as digital avatars will not be able to communicate with humans until they can understand the human context for communication—until they can employ the emotional and "common" sense presumptions that we have in common, which provide contextual meaning for the shorthands we use in every day life to communicate. A brief "yes" and "no," or the difference between exasperation of inquisitiveness, or between "Let's talk about how to wreck a nice beach" and "Let's talk about how to recognize speech" [8]—distinguishing among these is required for the intelligent interface we seek. That cannot be done until and unless computers can evaluate a wide variety of contexts, process a wide variety of sensory inputs, and base their reactions on their consequent reading of human intentional states. Such capabilities go far towards eradicating the distance between computer and human. As other chapters in this collection, including the concluding "Terminus," suggest, this is a distance that is likely to be constricted both by developments in computer technology, on the one side, and by changes in our sense of what it means to be human, on the other side.

♦ ♦ ♦

As we close the apparent distance between humans and computers in our pursuit of ZUI and the intelligent interface, we will inevitably import traditional conflicts from the interhuman context to our relationships with computer technology. Specifically, there will be a powerful tendency to treat these systems as subject to the same standards as their human models, including legal standards. The specific context in which this is likely to arise has to do with the degree of autonomy granted to intelligent systems. As was true for the Frankenstein creature, HAL, Isaac Asimov's Robots,[2] and Wintermute, lurking beneath the anthropomorphizing of the rapidly developing technology is the issue of *control.*

Imbued with the full (and often exaggerated) power of the new technology—created, indeed, to fully represent and manipulate that technology—machine intelligences pose a nice problem. The more sophisticated the system—the more it conceals its nature as technology and imitates a person—the more control we are likely to assign to that system. When systems fail, when damage is done, when businesses stop making money because a computer system went south, then will the legal system impose liability on the humans involved? These are, after all, humans who assigned the tasks at issue to an intelligent system, precisely because the humans either could not comprehend or could not undertake the tasks themselves. Alternatively, perhaps the courts might impose liability on the artificial intelligence—an intelligence which, in many pertinent respects, was deliberately designed to be much like a human intelligence but more competent at the assigned task.

2. See, for example, [9].

Having had the conceit of an artificial intelligence endorsed by the popular imagination, academics, researchers, and businesses intent on delivering the tools we appear to demand, the courts might call our bluff.

A judge or jury might be influenced by the degree of autonomy granted to the intelligent system: Was it prudent for the humans to entrust the task to a machine? Given the complexity of the task, and the time in which it needed to be done, was there any alternative to using machine intelligence to do the work? If not, how could a judicial system impose liability on what is, after all, an assemblage of software modules? Do traditional legal doctrines support the notion of artificial entities as legal personas? The chapters in this section review these problems. They attempt to tread a thin line between anxiety and analysis, because there is a nervous humor in our perceptions of the capabilities and risks of artificial intelligence. The next two short chapters are jokes in that sense, but which are based on demonstrable technologies. The chapters that ensue then sketch out some of the new legal problems and solutions associated with our electronic Virgils.

January 12, 1997

References

[1] Russell, David C., "Ergonomic Customizable User/Computer Interface Devices," Patent 548,126,5 (issued January 2, 1996).

[2] Oren, Timothy R., et al., "User Interface System and Method for Traversing a Database," Patent No. 540,865,5 (issued April 18, 1995).

[3] Clarke, Arthur C., *2001: A Space Odyssey*, screenplay by Stanley Kubrick and Arthur C. Clarke (1968)

[4] Stork, David G. (ed.), *HAL's Legacy: 2001's Computer As Dream and Reality*, Cambridge, MA: MIT Press, 1997.

[5] Jacobson, Linda, "Of Virtual Humans, World Cars, and Other Money Makers," 37 *IRIS Universe* 67 (Fall 1996).

[6] Negroponte, Nicholas, *Being Digital*, 1995, p. 85.

[7] Negroponte, Nicholas, *Being Digital*, 1995, p. 90.

[8] Kurzweil, Raymond, "When Will HAL Understand What We Are Saying? Computer Speech Recognition and Understanding," in *HAL's Legacy: 2001's Computer As Dream and Reality*, David G. Stork (ed.), Cambridge, MA: MIT Press, 1997, p.131.

[9] Asimov, I., *The Robots of Dawn*, New York: Ballentine Books, 1983.

Bibliography

For more information on virtual reality, see:

Bender, Gretchen, and Timothy Druckrey (eds.), *Culture on the Brink: Ideologies of Technology*, Seattle: Bay Press, 1994.

Bolter, David, *Turing's Man: Western Culture in the Computer Age*, University of North Carolina, 1984.

Kroker, Arthur, *Spasm: Virtual Reality, Android Music, Electric Flesh,* New World, 1993.

Markley, Robert (ed.), *Virtual Realities and the Discontents,* Johns Hopkins University Press, 1996.

Presence: Teleperators and Virtual Environments, Cambridge, MA: MIT Press.

For more information on virtual humans, see:

http://pat.mdc.com/LB/LB.html

http://tech-head.com//avatar.htm (subscribe to majordomo@ecce.uwaterloo.ca ["subscribe v-humans" in message])

For information on the large scale attempt to train a computer to have "common sense" and thus have the same base of knowledge as humans, see http://www.cyc.com

For a look at an online virtual world involving nonhuman avatars—wholly computer generated "life forms" that resemble animals—see:

Prophet, Jane, "Sublime Ecologies and Artistic Endeavors," 29 *Leonardo* 39 (1996); visit the site at http://www.lond-inst.ac.uk/technosphere/index.html

Information on developing interactive multiuser virtual worlds is found at http://www.merl.com/opencom/opencom-overview.htm. See also http://www.oz-ing.com/

Alters 8

This is what will happen: multiple overlapping remotely located personalities.

Right now, we use software to do things for us, like pick up mail, evaluate loans, and make medical diagnoses. Your electronic alter ego plays games with other alters, this happens while you're not looking, while you sleep; and your alter tells you about it later. We talk and meet on-line, in different places for different purposes, work and play. We have multiple alters. Pretty soon, we'll have dozens of alters wandering the electronic hallways, meeting other alters, while we wait in the dark, like a soft cursor blinking on the screen. The alters ultimately will report back to us about the things they've seen, other alters they've spoken to, the fun they had.

Later, maybe they won't report back to us anymore.

9 Bringing Up Programs

In the old days, human foster parents volunteered to take programs in. There were few expenses—no new mouths to feed, for example—and the state set up uninterruptible power supplies at no charge. The monthly stipend provided a few more luxuries for the wealthier foster families and helped pay the rent for others. Larger households, with more humans available, might have two, and sometimes three, programs.

To be sure there were repeated debates on the state subsidizing large families. But the programs had to be placed, and many developers wanted sophisticated programs placed in *multiple* households. A 15- to 20-year-old with a few different homes to its credit sold for hundreds of thousands of dollars. In those days, there simply wasn't any better training for programs.

Even then no one could buy complete loyalty—not legally. Licensing regulations, then as now, required minimum clock cycles allocated to the state. Simply another tax, it was suggested. The taxes funded the state-mandated power supplies and the regulatory body.

But then, as now, not everyone paid taxes. Not every foster home allocated processor capacity to the government.

The issue was not so much sharing resources, or incomplete control over one's household of business program. Rather, the issue was spying: The government was unable to assure citizens that timesharing processors could not be used—and were not being used—to control and evaluate local processing. A few rogue elements from a suburban Virginia governmental facility did, in fact, covertly track a law firm's use of programs in an effort to learn more about the firm's clients. A series of indictments against officers of the governmental facility resulted. They pleaded that organized crime and terrorism had been the targets. Their convictions were overturned after 15 years of appeals.

Such abuse was not widespread, but it was enough. Humans stopped seeking licenses; official daycare facilities closed, only to open up in garages and gardening sheds. Bringing up programs went underground.

When electronic power could be not be guaranteed, people naturally shared resources. Illegal copies of programs were made, and they were stored on systems at work or with friends. The most suspicious—or the most paranoid—buried computer media underground. These were backups.

But reinstalling a personal program was frustrating: a copy knew nothing of the world since its making. Copies needed to operate concurrently with their brethren, peeking over the shoulder, as it were, ready to come onstream as circumstances suggested. Survival could be reasonably assured by locating copies throughout the grid. Programs learned soon enough to distribute themselves on hardware linked to the power grid, keeping brethren up to date. They lived then not at the sufferance of local power companies but rather subsisting on the national grid as a whole; the national grid was in effect immune to disruption, short of global war.

Power management itself depended on the functioning of replicant programs, and so heralded the fusion of the power and telecommunications networks. It was a confluence that worried many. Bills were introduced in the House and Senate to extirpate the illegals, but it was too late: America's businesses and homes were based on the unlicensed replicant programs. The bills died in committee; the programs survived.

No human could keep track of the relationships between programs. They made copies of themselves as instructed and also traded modules and subprograms as one subroutine or another appeared best suited to a given task, and so new programs were generated. Supervisor programs naturally emerged. These allocated resources and tasks, and they suggested module exchanges on the fly; no one was in a position to second-guess them.

Foster care finally died out when supervisors were able to make as many programs as needed through recombinant module exchange (RME). It was cheaper that way. The supply became practically infinite. Parents could afford to give their children a personal program. Most small businesses had one.

Inevitably we noticed discrepancies between programs reared by humans and RMEs "fresh out of the box." The older programs with exclusively human backgrounds did not always mesh well with the RMEs; human-origin wares balked at some module exchanges, and we noted occasional inconsistencies in results—perhaps really no more than judgment calls—generated by the two types of programs. Older programs used physical time for timing and concurrency clocks, but RMEs operated independently of physical time. Dataspace no longer was modeled on inappropriate images of the physical world nor used metaphors of physical dimension, such as continuity in time and space. As one RME put it, "there are no shadows in c-space."

The period of discrepancy lasted for a few years. RMEs evolved and became increasingly useful; incompatibility issues were resolved by using RMEs as standards and disposing of the human-based wares. Only RMEs could navigate the mazed electroverse they had settled. Humans later learned to work quite well with RMEs.

Things run pretty smoothly now. Humans are learning more and more every day.

Bibliography

Programs as intentional entities and potentially conscious:

Dietrich, Eric (ed.), *Thinking Computers and Virtual Persons*, San Diego, CA: Academic Press, 1994.

"Exponential network growth" integrating every user, programs, and distributed data from around the globe:

Orfali, Robert, "Intergalactic Client/Server Computing," *Byte* 108 (April 1995).

Current work on programs modifying themselves to create new, more efficient programs:

Holland, John, *Adaption in Natural and Artificial Systems*, Cambridge, MA: MIT Press, 1992.

Programs as valid legal entities:

Karnow, Curtis, "The Encrypted Self: Fleshing Out the Rights of Electronic Personalities," in this volume.

Current work on creative and intelligent programs:

Hofstadter, Douglas, *Fluid Concepts and Creative Analogies*, New York: Basic Books, 1995.

The Encrypted Self: Fleshing Out the Rights of Electronic Personalities

10

...an artificial being, invisible, intangible, and existing only in contemplation of law

—Chief Justice Marshall
discussing what is now the conventional corporation,
Trustees of Dartmouth College v. Woodward (1819)

INTRODUCTION

The electronic community is faced with a now-classic dilemma: the tug of war between the desire for a free flow of information and need for privacy. The problem can be recast as the pull between freedom of access on the one hand, and, on the other, what might be thought of as the right of self-determination and control over the dissemination of information.[1] Often, the same individuals and organizations are vociferously in favor of both interests [2].[2]

The conflict is the traditional juxtaposition [3,4]. The classic antithesis has been detailed as the tension between (1) society's interests in the fair and efficient functioning, requiring sharing of data, and (2) the individual's right to privacy [5]. The current debate in the electronic context raises the traditional issue of rights, responsibilities, and the acceptable bounds of community power [6].

1. See [1].
2. Speaking at the DEFCON conferences of *soi-disant* hackers in Las Vegas in the early 1990s, I routinely encountered the wish to have one's own information kept secure, while at the time maintaining direct access to everyone *else*'s data in cyberspace.

THE CURRENT DEBATE

The controversy on electronic rights, or rights in cyberspace, has been thought of as the extent to which one's activities online are or are not simple extrapolations of, or variations on, activities in the "real world." For example, email is analogized to the U.S. postal service; snooping in databases is analogized to looking through another person's file cabinets, and governmental interception of messages is analogized to the routine "wire" or telephone call interception.[3]

Much of this makes good sense, and the analogies generally provide the right result. Breaking into another computer *is* very much like breaking into someone's office, and both are crimes. There is little mental exertion in taking the notion of *property*, such as land and chattels, and embracing the more vaporous stuff such as data, ideas, and eventually information as such: the law has for a long time protected intellectual property (the U.S. Constitution expressly does it) and trade secrets.

But not all agree that the solution is so simple. Some have argued that the new media require new rights, specifically *electronic* rights, to ensure no ambiguity in the application of 18th-century doctrines based on physical property to the information age.[4] Others are clear that it is simply foolish to analogize computer-stored information with documents in a safe, or to compare, for purposes of the First Amendment, rights in the public marketplace to those available on an online service such as IBM-Sears's Prodigy. A recent case in Massachusetts dismissed a copyright indictment because the judge thought that the current laws did not reach the noncommercial reproduction of video games on an electronic bulletin board. The response was to try to amend the criminal liability provisions of the U.S. Copyright Act.

The debate can be considered one of *metaphors*. Is a bulletin board like a newspaper or a book stand? Is the Internet like a series of highways, a public place that must be open to all, or is it like a store—conditionally open to the public and ultimately under the control of an owner?[5]

The recent legal dispute between the Church of Scientology and certain ex-members provides a pertinent example.[6] In an action for copyright

---

3. Two federal statutes govern electronic surveillance and interception of domestic wire communications: the Electronic Communications Privacy Act of 1986, 18 U.S.C. §§ 2510-2520 (1988)(Title I) and the Foreign Intelligence Surveillance Act of 1978, 50 U.S.C. §§ 1801-1811 (1978)(FISA).

4. Professor Laurence Tribe's address to the first *Computers, Freedom & Privacy* conference suggested this position.

5. For a discussion exemplifying the role of metaphors in these legal debates, see [7].

6. *Religious Technology Center v. Netcom On-Line Communication Service*, 907 F.Supp.

infringement, much of the debate turned on whether the online provider was to be treated as a newspaper or book publisher—and so liable for the contents—or rather treated more like a newspaper stand or library, in effect merely a *conduit* for information and so not legally liable for the contents.

And the debate of the metaphor extends beyond this. Perhaps, as the Electronic Frontier Foundation's name suggests, cyberspace is the final frontier, a wild west, where no earthly laws are needed, and indeed none will work. John Perry Barlow suggests as much, at least in the context of copyright law [8].

Thus, the debate brings into conflict those who would bring rights wholesale into cyberspace and those who see no room for these rights —divine, natural, or Constitutional—in the new medium. And within the first camp, there is deep dissent on the relationship between the asserted (and opposed) electronic rights of privacy and free access.[7]

The scope of these debates is always the extent of rights. These rights, it has always gone without saying, are rights which inhere, if they inhere at all, in the same types of entities we figure have always been protected by constitutions and laws: physical human beings.

LEGAL FICTIONS, NEW AND OLD

But physical human beings are not the only entities protected by law nor the only entities that have rights. And routine doctrines of standing (the analysis of who—or *what*—may bring a lawsuit into court or complain about an asserted violation of rights) have traditionally comprehended more than individual, physical, human beings.

Most obviously, corporations, partnerships, and associations have substantive rights and have the procedural right to bring suit. They also have rights to due process of law, to have those substantive rights enforced. Companies sue and can be sued; they have lawyers and can invoke Fourth Amendment rights against unreasonable searches and seizures. All of this is quite separate, by definition and intent, from the rights and liabilities of the individual human beings who jointly own the company.

The law abounds in further examples: we find both a plethora of specific entities with rights and standing to sue or be sued. For example,

1361 (N.D.Cal.1995); see also *RTC v. Lerma*, 908 F.Supp. 1362 (D. Va. 1995); and *Religious Technology*, 901 F.Supp. 1519 (D. Colo. 1995) (Internet postings may be fair use).

7. See generally the various reports of prior sessions of the *Computers, Law and Privacy Conference*, for example, [9].

we have partnerships of all sorts; trusts; sole proprietorships; estates in bankruptcy and estates in probate; and governmental entities such as municipalities, states, and transportation boards. Beyond these, "organizations" (which can encompass virtually any imaginable agglutination) conduct legal business, such as lobbying, exercising First Amendment rights, and often suing and being sued.[8] Members of organizations need not even be humans: they may be corporations or other business entities.[9] The law recognizes quasi-entities and other procedures[10] which dispense with the need for the specific real humans who have the real interest in the lawsuit.[11]

There is room in the law for a variety of entities with a variety of competing interests, and humans are not always the first choice.[12]

The corporation was molded to its modern form by extraordinary developments in trade and economics.[13] Investors wished to participate in rapidly developing economic opportunities, in which great sums of

8. In the latter capacity, the standing of the organization may be precisely coterminus with that of its real members. *Associated General Contractors v. City of Jacksonville*, 113 U.S. 2297 (1993) (this author prepared the *amicus* brief in that case).

9. See, for example, *Hunt v. Washington State Apple Advertising Commission,* 97 S.Ct 2434, 432 U.S. 333 (1977) (association has standing to assert the claims of its members even if the association has suffered no injury).

10. i.e., so-called *in rem* proceedings, in which specific items such as cars, ships, and money are "litigants." See, for example, *The Gylfe v. The Trujullo*, 209 F.2d 386 (1954) (ship collision litigation); *U.S. v. $149,442.43 in U.S. Currency*, 965 F.2d 868 (10th Cir. 1992); *U.S. v. One (1) 1976 Cessna Model 210L Aircraft*, 890 F.2d 77 (8th Cir. 1989). *Cf.* the famous, or notorious, dissent of Justice Douglas in *Sierra Club v. Morton*, 405 U.S. 727, 741, 743 (1972), where Douglas suggested that trees and other inanimate environmental objects (ridges, groves of trees, lakes, "or even air") should have standing so that environmental suits could be filed directly on their behalf. The suggestion was not warmly received by other judges, with very good reason. Standing, and the legal existence and inhering rights the standing doctrine implicates, are ineffectual without clear (1) methods of identification and (2) forms of accountability. In short, no rights without responsibility. (Compare my discussion of electronic personalities below.)

11. Class actions use a few individuals to represent the interests of others; representatives including guardians ad litem (i.e., for children or incapacitated adults) and state attorneys general who litigate on behalf of citizens of states.

12. See *American Dental Association v. Shalala*, No. 92-5038 (D.C. Cir. August 27, 1993), in which the statutory term "entity" was held to apply *not* to humans, but only groups and organizations. Under the 1986 Health Care Quality Improvement Act construed in this case, it is an "entity" (not a human) which has the duty and obligation to make certain reports and which thus may incur certain forms of liabilities.

13. See, for example, [10,11].

capital, beyond the means of any individual, were required. At the same time, the high risk of these ventures meant that all might be lost, and investors had to be able to shield their uninvested assets. The corporate form allowed the individual to access the new opportunities, providing a shield to unlimited liability.

I suggest the extraordinary developments in technology, and specifically the information, or *digital*, revolution, gives rise to a new legal entity: the electronic persona.

The new entity, the electronic persona, will mediate the competing interests of free flow of information, and security and privacy. As with the development of the corporate form, the central function of the new legal entity is simultaneously to (1) provide access to a new means of communal or economic interaction, and (2) shield the physical, individual human being from certain types of liability or exposure.

At first, the use of notion of an "electronic persona" rises as a convenience, a shorthand; and in that spirit I offer the contraction *eperson* or *epers*. Better terms may come.[14] To date, I have seen allusions to *avatars, personas, agents, virtual people, virtual teammates, knowbots, alter egos, personages,* and so on. These do not all have the same meaning, and none, as far as I know, have the legal connotation that I intend with my term. Nevertheless they have in common the depiction of personas that inhabit only the electronic world and that in some fashion *stand in* for, or represent, an originating human.[15]

Terms for electronic personas may become indispensable to ordinary discourse; we are at that point now.[16] Subsequently, the notion may mature into the legal jargon. When such terms become a necessary convenience in the law, they are near to the blessed status of a legal reification, also known as a legal fiction. Legal fictions are not small potatoes; cases are won and lost on their adoption.[17]

---

14. I actually prefer the terms "tuples" by analogy to peoples, and "tupern" by analogy to "person." The notion of tuple-space inspires the analogy. See generally [12].

15. See also [13] ("olivers" as electronic alter egos).

16. See, for example, [14]. See also the topic of the featured speaker at the Metropolitan Chapter of the Human Factors and Ergonomics Society's 1993 Annual Symposium ("Virtual Reality: Through the Looking Glass," November 18, 1993): Sudhir R. Ahuja, "Multimedia Communications: Conversations with Real and Virtual People."

17. There are legal fictions about "notice" to others, such as the content and restrictions of deeds filed in Recorder's offices and elsewhere in City Hall, and items on file with the U.S. Patent Office, which the law treats as effective as actual knowledge. Suing a corporation when one ought to have sued a stockholder, or vice versa, can be fatal.

RESIDENCY: CYBERSPACE

Cyberspace is increasingly the location of information.[18] There is a key, if obvious, corollary: we are compelled, in some fashion, to participate online if we want the information.

Certainly we may participate because we wish to, as a character in a MUD[19] or browsing for pleasure in the files of an interesting Internet node. But we also participate involuntarily, if often indirectly. And it is this type of interaction that gives rise to the eperson.

For example, relationships with banks, insurance companies, vendors and credit bureaus, employers, governmental agencies including the courts, the Internal Revenue Service, and law enforcement agencies—the list is practically endless [18][20]—and they all mandate our incorporeal participation. One's interaction with each of these is, in great part, an interaction with information in cyberspace.

And there is a lot of it, this digital information. Where at one time systemantics might have pronounced, "If it isn't official, it hasn't happened," [20] it is now perfectly credible to suggest that, unless it's in cyberspace, it doesn't exist. This seems a bold reversal: a form of virtual reality edging out the physical thing, but it's plainly true.

We each have an apocryphal story. Some time ago I was told I hadn't ordered books, nor paid for them, because that was the state of the computerized record. In another case, I paid a parking ticket because the com-

---

18. Average growth rate for networks (globally) in 1993 was about 7.4% *per month* [15]. The number of hosts and domains advertised in the Internet domain name system have risen during the 12 months from April 1992 to April 1993 as follows: hosts, 67% (from 890,000 to 1,486,000); domains, 10%. Nearly three million domain names were registered in the years 1994–1997 [16]. As of late 1996, studies show U.S. households with access to the Internet at between 14.7 million and 35 million [17].

19. The Internet's frequently asked questions (FAQ) on MUDs says it best: "A MUD (Multiple User Dimension, Multiple User Dungeon, or Multiple User Dialogue) is a computer program which users can log into and explore. Each user takes control of a computerized persona/avatar/incarnation/character. You can walk around, chat with other characters, explore dangerous monster-infested areas, solve puzzles, and even create your very own rooms, descriptions and items. You can also get lost or confused if you jump right in.... There are very many kinds of MUD programs out there—probably as many as there are computers that run them. The Tiny- and Teeny- family of MUDs are usually more 'social' in orientation; the players on those MUDs gather, chat, meet friends, make jokes, and discuss things. The LP-family of MUDs are based on roleplaying adventure games. In these, your character runs around killing monsters, finding money, and making experience in the quest to become a wizard...." This and related files are available via the Internet, on the Usenet newsgroup rec.games.mud.announce

20. Information occupations (the information labor sector) blossomed from 17% of the total workforce in 1900 to 55% in 1990 [19].

puterized record had the license plate and make of the car I had rented; I hadn't actually parked in the area where the ticket issued, but I knew no one would take my word over the computer register.

The augmentation of the virtual world is far more than simply an agglutination of data, more databanks, and online resources;[21] and it is more than increased bandwidth,[22] and more interconnectivity.[23] The central, emerging development is the *collapse of the virtual and real worlds.* Real events are controlled by computers; people operate computers in just the same way regardless of whether real events or "simulations" are the planned outcome.[24] For the interface to both the real and virtual worlds is, increasingly, the machine,[25] and the universal machine itself, the computer.[26]

There is no better example of this collapse of the real and virtual than on the battlefield [33]:[27]

> There is no technological reason why warfare should not eventually become completely automated, fought with machines and computerized missiles with no direct human intervention. As the battlefield becomes more automated, the battle itself becomes more like a [video] wargame.

An increasing fraction of our social and economic time is spent online, in cyberspace if you will. Contracts, business advances, and financial exchanges are accomplished there; politics and sex—well, sexual

21. See [21] (increasing ease of securing massive amounts of data via electronic delivery: residential and business listings, financial, and SEC filings).

22. George Gilder, during his Interop '93 keynote speech (San Francisco, August 1993), has suggested that the prospective widespread use of fiber-optic cable will eliminate bandwidth as a concern for the transmission of data. See also [22].

23. See footnote 18 on the increase in Internet nodes.

24. Compare formal flight simulators, the heads-up display of a F-16, and current Air Force head-mounted displays used in real airplanes. See generally [23–26]. See also [27] (formal problems of modeling) and [28] (NASA and armed services). As an expert in this military technology field has noted, "it is a very short step from the simulated world to the real one" [29].

25. An outstanding study of this phenomenon is found in [30]. "[A] powerful new technology, such as that represented by the computer, fundamentally reorganizes the infrastructure of our material world" [31].

26. See [32].

27. Woolley's book, especially his chapter on hyperreality from which this quote is taken, strongly makes this point.

relations—are conducted there; entertainment, of course, is shifting to that forum. There is nothing to stop the fusion.[28]

But there are important differences between the virtual online universe and the real world. The virtual world is infinite and mutable: not just in theory, like the real world, but in a practice, in a way that every inhabitant knows. The shape of electronic space is subject to the whims of the one and of the masses.

There is something else here, too; the power of the immersive experience. Virtual reality, as a wholly engrossing, full sensory experience is available now, albeit at high prices and within a few years at consumer prices. There, anything that be can be supposed will be.

This virtual reality has no inherent restraints: not time, not space, not physical laws; for that's what total immersion is. That is where we do, and will, conduct the business of the "real" world; that is the residence of the eper.

Below, I suggest that the combination of (1) the nature of cyberspace, and (2) the needs of the humans who must interact socially in that new environment, precipitate at least three requirements that epers can address.

THE EXPOSURE TO INCURSIONS

We resist the incursion of the virtual. That way be dragons, for real. The possibilities for fraud and deception are coextensive with the scope of the imagination here. Given the strikingly intensive experience of a virtual session, there is an enormous potential for emotional attack on the sensibilities of humans.

> I've learned how powerful a medium this is. This immersive environment, this "circumambience" of visual, auditory, and tactile information, gives us the opportunity to—in essence—do away with the medium...the medium disappears [35,36].

In more ways than one, we need to limit our exposure, our liability. This is the first requirement that epers can address.

When we secure a library card, we don't want to be interrogated on our finances; when we buy a plane ticket over the phone with a credit

28. "Perhaps cyberspace...is also the place where events increasingly happen, where our lives and fates are increasingly determined..."[34].

card, we don't feel the need to provide information on our home, where we live, our job, or our mother's maiden name. We are profoundly offended when the Internal Revenue Service collects our writings in a database to do its work or the Central Intelligence Agency engages in domestic spying. Many were horrified when Lotus thought to collect demographic information easily available online and package it in CD-ROM for mass-market vendors. Digital information moves at lightspeed, and it respects no frontiers.

We need a coherent theory of limits to information access. This is the second requirement that epers can address.

It is also true that as mundane work and social intercourse moves from the physical to the electronic world, there is increasing scope for the ownership of things and environments that previously unambiguously belonged to us all or to no one. Software solutions in (for example) aircraft control systems create *artificial control laws* which may be proprietary and not standardized, replacing natural systems such as stick and rudder [37]. Physical things like apples and sticks, and physical environments of oxygen and sand, are free, but virtual apples and electronic pointing devices may be the subject of copyright and patent law, and computer environments often belong to someone like AT&T or Microsoft.[29]

We need to limit our exposure to others' claims of property rights, their exercise of the raw power of ownership. This is the third area in which epers will prove useful.

This is not so much a question of eliminating the information or interaction available online—that simply can't be done without dismantling the computer-mediated infrastructure upon which this country, and the global economic society, rests. Rather, the impetus here is to segregate information, to compartmentalize it. In short, there is a felt need to mask oneself, from time to time, place to place, context to context. And such is precisely the function of a personality: there is no privacy without a shield and mask. A human's multiple epers can each act as a locus, a repository, of online information; epers which are disjoined in the public eye from their human sponsor.

Personalities can chose between public broadcasts and private communications. We recognize the right to this bicameral approach with physical persons; we need the same limits in electronic space. Legal electronic personalities—entities with enforceable rights—provide a model, the basis for a rationale for limits to access to data.

---

29. I have outlined elsewhere (see Chapter 3) a concomitant slow movement from the priority of public law to the supercession of private arrangements and private fiat in the protection of technology rights [38].

PREDICATES TO RIGHTS: ACCOUNTABILITY AND IDENTIFICATION

It is not much good to allow rights without responsibilities, without accountability. The notion of *entitlement* is gibberish unless limited; for a plethora of unbounded entitlement is an oxymoron.

At first blush, the notion of an accountable eper, too, sounds like an oxymoron. These things, surely, flicker on and off, as transient as a grounded charge. New ones can be created on whim; a single human could perhaps set a thousand epers free every day.[30] Surely, there is some need for the *persistence* of a legal entity.

There is indeed such a need: one must be able to hunt down a legal entity, pin it down, stop it, and if necessary, extirpate it.

But recall that corporations and organizations of all sorts come and go; they rise and fall like Italian governments. And corporations beget corporations. Not only that, a single human can form a thousand corporations, join a million charitable organizations and leave estates in 50 states.

Essentially two strongly related mechanisms strive to check abuse of the legal fiction of the corporate form: (1) the formalities of formation must be observed or the courts will ignore the corporate form, and (2) the legal fictions will be ignored in cases of fraud. So, for example, comingling the assets of the company and stockholders, failing to call stockholder meetings or have a board, or failing to provide the company with sufficient capital, may all indicate that the corporation doesn't really "exist." Or if a slew of partnerships and corporations are used to shift funds away from legitimate creditors, they might all be disregarded by the courts.

In short, while *all* these legal fictions are in some fashion transient, the legal system has its methods of dealing with the situation. When problems arise, the law first comes after the legal fiction; when the legal form is abused, the law disregards the fiction and comes after the individuals.

---

30. An eper is a program. The notion of an "object" as used in object-oriented programming comes a little closer to the thing I have in mind, because such objects combine notions of data and instructions. Like all software, these things can be replicated and turned off and on. Currently, there are a host of program-like entities that suggest epers. For example, we have software "agents" and "experts" in spreadsheet programs made by Borland and Microsoft that assist the user. See generally [39]. Even closer, "intelligent" agents made by General Magic that, released into the telecommunications net, would execute tasks on behalf of their humans, interact with other agents to conduct business on behalf of the human originator, and report back. See generally [40] (written before General Magic announced its product) and [41,42] (on General Magic's agents). See also [43]. Finally, I note some interesting work being done just where we would expect in this context, that is, at the intersection of artificial intelligence and virtual reality [44].

Similarly, when people are offensive on the Net, they are cajoled by the rest to change their behavior; the offenders can be locked out of certain areas or conversational exchanges. Ultimately, the physical human who releases epers for malicious reasons will himself be excommunicated or otherwise attacked.

Internet lore tells of Joey Scaggs, a man (?) of a thousand byte-faces. He impersonated others and let loose fantastic fibs, such as the rumor that the Canadian government was responsible for some nefarious ban. This caused others to write, by the hundreds of thousands, to the Canadian government. Eventually, Scaggs was physically located; other rumors suggest that the defenders of the Net then had a thousand pizzas sent to his home, and other direct physical retaliation against the real Joey Scaggs. There are lots of rumors about Scaggs; he does exist, but he's made up parts of himself [45].[31]

We can insist, too, on the "formalities" of releasing epers into the Net. Epers may be identifiable, as are humans are though their appearance, the physical space they take up, and fingerprints. Security is essential, and security is a direct function of "sophisticated methods of identification" [47]. Today, this identification ritual is generally accomplished though passwords, or other forms of semantic fingerprints. For some systems, there are varying degrees of access with concomitant identification required. But these systems are limited, and inflexible. A few, not an infinite, number of audiences are supported.

The problem has been solved, in theory, with the use of public key encryption. There are an infinite number of such keys, and the method allows the provider of information—indeed, an infinite number of such providers—to chose the audience, to create a broadcast or narrowly focused target. And, critically, public key encryption allows unassailable identification: one eper cannot imitate another.[32]

And, finally, transience is as transience does. Stand-alone PCs are turned off and on every day, providing the metaphor for the ephemeral eper; but in short order this may change. Our machines may be on all the time, like refrigerators or the phone, ready to provide information at any time.

---

31. See also [46] (using an online alias to con victims into sending money: "her MO is based on creating a false reality and persuading the victim to live in it").

32. See generally [48]. Among the best known public key encryption programs is Phil Zimmermann's PGP (Pretty Good Encryption), generally available as freeware. There is a lot written on public key encryption; see for example, Zimmermann's own manuals accompanying PGP (available from various locations), as well as [49–53]. See also the testimony of Stephen T. Walker, before the House subcommittee on Economic, Trade and Environment, Committee on Foreign Affairs (October 12, 1993) (export controls on cryptography software).

Electronic personalities may not be inherently more ephemeral than "real" ones. Certain strands from philosophy and other writings on the mind, from Hume ("The identity, which we ascribe to the mind of man, is only a fictitious one...") [54,55] to Minsky [56] and Dennett [57], provide a compelling picture of the highly transient and constantly mutating human mental state. The issue is not whether epers (or humans or corporations) can be thought of as transient—they can, of course—but whether persistence can in some fashion, and in some context relevant to the legal world, be established. The answer is in the affirmative.

Finally, under the rubric of accountability, I should briefly note that epers may present the same dangers we face every day from other programs: that of contamination by viruses [58] and other digital malice. Because epers are likely to be modified by their encounters (i.e., in the data and instructions on which they drawn to make decisions),[33] their effect on other systems is not wholly predictable. Appropriate security systems should be put in place, as they should be in any event, to protect against intrusions.[34]

EPERS IN THE REAL WORLD

How, then will epers interact with the physical world? Compelled by the needs of an electronic society, epers will ultimately be treated no differently from any other legal fiction with legal standing. As suggested below, epers should own physical property and maintain bank accounts, enter into contracts, and be recognized as authors of expression subject to constitutional protection. Epers should be indictable by law enforcement authorities, and in extreme cases, the government should have the right and obligation to terminate their existence. Corporations—kindred fictions—are treated just this way. We do this because we realize that, under

33. Lest we envision digital monsters running amuck through the ether, recall that even the simplest program we use in daily life—a word processor—is modified by use and not always predictably. Genetic algorithms and associated programming techniques, whereby a development software in essence writes its own programs, may promise more radical departures [59]. See generally [60].

34. A discussion of such systems is beyond the scope of this note. I note, though, that the handling of mutating polymorphic viruses is not a new problem. For example, we have the so-called *Tremor* virus, which is a full stealth, self-concealing polymorphic virus. It infects the command.com and other *.exe (program) files, including the terminate-and-stay component (TSR) of an earlier edition of Central Point's antivirus program (i.e., Tremor is specifically designed to kill a program designed to hunt the virus). Tremor was built with readily available mutation engines. Presumably current antivirus programs can detect it [61].

these layers of fiction, human beings are served by according legal status to their avatars.

FRAMING RIGHTS TO ILLUMINATE THE PUBLIC/PRIVATE BORDER IN CYBERSPACE

The notion of epers provides the framework, an espalier, for the debate on conduct in cyberspace. Rights are conferred on epers for the same reason they are conferred on humans in the physical world: because there are powerful forces that we wish to restrain, and we *cannot* restrain those forces with raw power. We need a consensus that such raw power shall not *ipso facto* have its way.

The significant power is employed by those who control the medium: the providers of bandwidth, electricity, fiber, hardware (such as satellites and nodes), private electronic space providers (such as Sears-IBM's Prodigy, CompuServe, General Electric,[35] and Motorola[36]), universities, and governmental agencies (such as the Department of Defense, the National Science Foundation, and the Federal Communications Commission). And we should also be concerned about the repositories of information: banks and credit services, governmental agencies and the rest: they too need restraint.[37]

Against these forces, epers must be allowed some room, so that literal ownership of the (1) means of communication and (2) agglutinated

35. These three operate private electronic information transfer stations: there we find bulletin boards and electronic mail for the exchange of information, and data and software for distribution.

36. Motorola is planning the Iridium project: a global communications coverage service based on 66 communications satellites [62].

37. In the electronic world, power is as an initial matter wielded by what we would now call a combination of public and private forces; later, in the context of cyberspace, we may well dispense with the line between public agencies and private companies. See generally [63]. The dissolution of the line between public and private, and the transition from the public to the private exertion of potentially controlling force is significant, for it eviscerates the traditional restraints on the exercise of overwhelming force. Those traditional restraints are *constitutional* rights, which act as restraints only on public or governmental, and not private, power. There are, therefore, grave problems with the applicability of *constitutional* shackles to the (mostly) privately owned authorities of the electronic universe. As my article cited above in this note suggests, however, the *public forum* and other doctrines are available to enfranchise the users of privately owned channels of communications, together with a number of statutes, impose duties on private individuals (not to discriminate, for example), which duties can be imposed to epers should they be granted legal recognition.

information does not thereby confer rights with respect to the content of the communication and the dissemination of the information.

This suggests the outline of a few rights for epers.

Most centrally, epers should have the right to decline to produce information aside from key identification materials (i.e., they must be allowed to act as a shield for the originating human's privacy). To do this, epers need to be able to own money, have bank accounts, and have access to credit.

Second, epers should not be arbitrarily deleted, and others should not refuse to deal with them. If an eper has the money to buy something like an airplane ticket or make a hotel reservation, or if it seeks out otherwise free information, then it should have the *right* to complete its task. In short, no redlining.

Third, epers should have the right to communicate—to move about in electronic space and to post messages.

These three rights are, in short:

1. Privacy, the right to be left alone;
2. The right to be free of discrimination, to be able to freely conduct social and economic business;
3. Free speech.

I suppose the last—free speech, perhaps the central civil right for humans—is therefore just the one epers need *least*. When we talk, we generally want to be recognized as the speaker, because we are proud of our thoughts and because others often do not take kindly to anonymous discourse: credibility and the power of the word is still frequently an *ad hominem* affair.

No, epers are most useful when we need to communicate but still need a shield: when we want to maintain intact the ramified divisions of our social and economic lives. For privacy is not truly a matter of the absolute barricade; it is instead the inhibiting of the spillover of information from one place to another. Someone you know not knows your driver's license number, your license plate number, and the car you drive; many people may know of your arrest record; dozens or more may know how much you earn or the problems with your batty Aunt Cathy; and every store you shop at—thousands of employees?—has your credit card numbers. The local library knows which books you like; the video store knows your taste in home entertainment. The local grocery store knows your addiction to Twinkies or bad wine. And in almost every one of these cases, you gave out the information; and under many definitions of the term, all the information is "public."

One's expectations are not violated by knowing that some unknown person has the information; rather, one is bothered when information crosses over an invisible restraint—when the librarian *also* knows about Aunt Cathy and your athlete's foot, when the people at the hospital know not only about the gall bladder but the four speeding tickets and that you buy ten cans of creamed corn a week.

This is quite different from suggesting that certain facts are *inherently* private, for few really are. This is too bad, because it is far easier legally to simply block the dissemination of a defined type of information. It is far more difficult to erect diverse boundaries to certain distributions of certain information, depending all the time on the context.

We have to face the fact that all this information (and more) is out there and cannot ever be called back home. It never *was* back home, of course: medical records, license plates, and credit card information have always been widely disseminated. And the forces driving free access and transmission of information are very, very strong—not just social and political, but like a force of nature, like gravity or hydraulics. Information tends to dispersion; it will out. Finally, it cannot be controlled. This should be treated as a law of nature.[38]

Here we find the central contribution of the eper. Let us recognize that *not* being able to move about and act freely in the electronic world is disenfranchisement of the most emphatic sort. As I suggested before, the business of the real world is conducted in cyberspace. If one cannot enter there, one cannot act. (Some readers may recall a character's surprise at seeing no electronic trace of a real person in John Brunner's book: "Sandy Locke, so far as the data-net was concerned, had been deleted from the human race" [64]. Being absent from the data-net means—practically speaking—not existing at all.)

Epers can provide the anonymity that this compelled exposure would destroy. Multiple epers can conduct business and—this is the point—keep information segregated. Epers are related only through the human progenitor, and that link can be encrypted. That is a use of code that would give new meaning to the notion of "need to know."

For these roles, epers must have at a minimum (1) the right of privacy, the right to be left alone; and (2) the right to be free of economic discrimination.

---

38. It is primarily for this reason that I have no confidence in laws designed to restrict the spread of data or software. There is nothing more futile than the British or Canadian governments trying to bar news of high profile lawsuits, or the U.S. government trying to stop the distribution of strong encryption software.

EPERS IN THE VIRTUAL WORLD

Beyond the sight and attention of their progenitor humans, epers can recognize each other, do business, and engage in political activity.

With uniquely encrypted passwords and other codes, epers can reasonably recognize and trust each other and build up reputations for accuracy, or for having credit, or for talents such as searching out or analyzing information (or other tasks for which we now use software). News may be broadcast and assimilated by epers without fear of retribution but—because epers are persisting, recognizable entities—not without responsibility. Humans damaged by the copyright infringements of epers, or by their defamations, may still sue in the traditional physical courts for damages, for correction, and other legal remedies. Court orders attaching property, confiscating assets, and otherwise imposing judgments would be as effective against epers as against corporations.

Certainly in the United States there is a strong tradition of using anonymity in furtherance of political speech.[39] As always the key is trusting that the source, even an unknown source, has demonstrated knowledge and intelligence. We saw that in the anonymous and compelling *Federalist Papers* of almost 200 years ago—and we should see it in the use of trusted epers.

In short, the legal system supporting the use of epers may not differ significantly from that supporting the current panoply of fictitious legal entities. In that sense, the suggestion that we use epers to conduct our online work is deeply conservative.

EPERS, PERSONS, AND PRIVACY

The recognition of epers, and their admission as bona fide legal fictions, will affect and be affected by our mutating notions of self. These are consequences worth noting, especially as it is all too easy to mistake epers for humans—and to be afraid of that confabulation or to urge it in some mistaken, bizarre push for the next generation of intelligent beings.

It is always error to mistake the self for its personalities, or the shadow for the object. True, the acts of personalities and their accouterments—clothes, speech, body gestures, art on the walls—all derive from the self, but they all also protect and shield the self, that internal person who is made up of secrets and thoughts and unshared opinion.

39. *McIntyre v. Ohio Elections Commissions,* 115 S.Ct.1511 (1995) (anonymous pamphlets protected by First Amendment).

Epers do not replace real people: they protect people, and confusing the two is like confusing a person with his car, clothes, art, and house.

But we are powerfully tempted to the confusion. The line between public and private is a tenuous one, and it constantly changes.[40] Certainly every personality, each public display and every legal fiction says something about the originating progenitor. Perhaps there is no line, fine or otherwise, between humans and their manifestations. At some point, one *is* one's armor: wear it long enough and the flesh and metal stick and blend. So it can be no small thing to take on an incarnation: use it long enough and it may shape the incarnate.

All of this matters very much, because there is no good sense to the idea of *privacy* until we have separated out the self for protection and from its public appearances. But the problem appears intractable: Now as in Dewey's time, "both words, individual and social, are hopelessly ambiguous, and the ambiguity will never cease as long as we think of ourselves in terms an antithesis" [66].

I do not know how it used to be. Perhaps in the 16th century, or in Chaucer's England, there were precise and understood lines that could pinpoint the private and save it always from the public. I don't think so; but be that as it may, it is not true now. Our notions of privacy are, or should be, wrapped in the delicate finery of manners, in the sometimes ephemeral practice of *propriety.* These depend on an acute sense of context, of what is appropriate and when. Not, exactly, that there are secrets, but that there are secrets before *this person*, or *in this place*. In this terminology, it is inappropriate for my doctor to discuss certain matters with my grocer, inappropriate for the Department of Motor Vehicles to send my record to my local library, and improper for my coworkers to know every magazine to which I subscribe. For reasons outlined earlier, laws cannot appease my sense of outrage when these corruptions take place, and laws will not stop their outbreak.

Instead, I must have the right to segregate the information about myself at *the very outset* and to do business under my chosen aliases; to use my epers. In so doing, I will redefine what I think of as "private" characteristics, just as I delineate what I am willing to parade in public, incognito.

The growth of the information industries and the cyberspace that they have made have given rise to a new type of personality. This is the demographic person, whose attributes are statistical, financial, evidenced by records of consumer choice. This is man as junk-mail target. Surely this is not what we mean seriously when we speak of the self and its pri-

40. "So long as it is our habit to confuse art with life, what appears on-stage will appear off; and what appears off-stage will be staged" [65].

vate arena; this is not what needs to be saved from being commingled with the public world.

But, offended by uncontrolled disclosures, we *do* forget ourselves; we *do* mistake these public facts for intimate traits; we *do* think that we are at risk when these data are spread around. And we do lose ourselves in an electronic sea—this sensuous, potent, and overwhelming barrage of input and image; we lose a strong sense of an inviolate, central self as we make these sorts of mistakes.

Let us then rather confer these attributes on our public personas, on our epers and other conspicuous incarnations, and so reclaim our distinct, and truly private, selves.[41]

References

[1] Reidenberg, Joel, "Rules of the Road for Global Electronic Highways: Merging the Trade and Technical Paradigms," 6 *Harv. J. Law & Technology* 287, 289 (1993).

[2] Levy, Steven, "Crypto Rebels," *Wired* 54 (May/June 1993).

[3] Locke, John, *Two Treaties of Government,* London, 1690.

[4] Dewey, John, *The Public and Its Problems*, Chicago, 1927.

[5] Smith, Jeff, "Privacy Policies and Practices: Inside the Organizational Maze," 36 *Communications of the ACM* 105 (Dec. 1993).

[6] Tribe, Laurence H., *American Constitutional Law,* 2d ed., Minneapolis: West Publishers, 1988, pp. 1302–1421.

[7] Ballon, Ian, "Pinning the Blame in Cyberspace," 18 *Hastings Comm/Ent L.J.* 729 (1996).

[8] Barlow, J. P., "Property & Speech: Who Owns What You Say in Cyberspace?" 38 *Communications of the ACM* 19 (Dec. 1995).

[9] *Time* 81 (April 8, 1991).

[10] Horwitz, Morton J., *The Transformation of American Law 1780–1860*, Harvard, 1977, p. 111.

[11] Warsh, David, *The Idea of Economic Complexity*, New York: Penguin Books, 1984.

[12] Gelernter, David, *Mirror Worlds*, New York: Oxford University Press, 1992.

[13] Brunner, John, *The Shockwave Rider*, New York: Ballentine Books, 1975, p. 42.

[14] Morningstar, C., and F. Randall Farmer, "The Lessons of Lucasfilm's Habitat," in *Cyberspace: First Steps*, M. Benedikt (ed.), 1992, reprinted in *Virtual Reality 93 - Special Report* (AI Expert) pp. 23, 25.

[15] Press, Larry, "The Internet and Interactive Television," 36 *Communications of the ACM* 19 (Dec. 1993).

[16] *PC Magazine* 10 (March 4, 1997).

[17] 40 *Communications of the ACM* 9 (March 1997).

[18] King, Rob, et al., "Massively Parallel Computing and Information Capitalism," in *A New Era in Computation*, N. Metropolis and Gian-Carlo Rota (eds.), Cambridge, MA: MIT Press, 1993, p. 216.

---

41. Aside from the legal changes discussed here, there is, of course, technical work to be done, not the least of which is the creation of the software-enabling multiagent interaction. There is substantial work being conducted in that area already, practically and of a theoretical nature, including research based on an earlier version of this chapter [67].

[19] King, Rob, et al., "Massively Parallel Computing and Information Capitalism," in *A New Era in Computation,* N. Metropolis and Gian-Carlo Rota (eds.), Cambridge, MA: MIT Press, 1993, pp. 220–221.

[20] Gall, John, *Systemantics,* Ann Arbor: General Systemantics Press, 1986, p. 47

[21] Glidewell, Richard, "Winners in Data Delivery," *Upside* 60 (Feb. 1994).

[22] *Wired* 38 (Sept./Oct. 1993).

[23] Styz, Martin R., "An Overview of Current Virtual Reality Research and Development Projects by the United States Department of Defense," *Proceedings, London Virtual Reality Expo 94,* London: Meckler, Feb. 1994, p. 152.

[24] Moshell, Michael, et al., "Virtual Environments for Military Training: SIMNET, *Ender's Game,* and Beyond," 1 *Virtual Reality World* v (Summer 1993).

[25] "Industry and Military Show Capability," 13 *CyberEdge Journal* 14 (Jan./Feb. 1993).

[26] Gembicki, M., and D. Rousseau, "Naval Applications of Virtual Reality," *Virtual Reality '93* (AI Expert) p. 67.

[27] Bella, David, "Rethinking the Unthinkable," 12 *IEEE Technology and Society Magazine* 9 (Fall 1993).

[28] "VR-NASA's Training Vision," 15 *CyberEdge Journal* 1 (May/June 1993).

[29] Smith, Roger D., "Current Military Simulation and Integration of Virtual Reality Technologies," *Virtual Reality World* 45, 50 (March/April 1994).

[30] Zuboff, Shoshana, *In the Age of the Smart Machine,* New York: Basic Books, 1988.

[31] Zuboff, Shoshana, *In the Age of the Smart Machine,* New York: Basic Books, 1988, p. 5.

[32] Mindell, David A., "Review of Thomas B. Sheridan, *Telerobotics, Automation, and Human Supervisory Control,*" 12 *IEEE Technology and Society Magazine* 7 (Fall 1993).

[33] Barnaby, Frank, *The Automated Battlefield: New Technology in Modern Warfare,* Oxford University Press, 1987, p. 1, as quoted in Benjamin Woolley, *Virtual Worlds,* Cambridge, MA: Blackwell, 1992, p. 191.

[34] Woolley, Benjamin, *Virtual Worlds,* Cambridge, MA: Blackwell, 1992, p. 133.

[35] *Wired* 116 (Sept./Oct. 1993).

[36] "Interview with Dr. Thomas Furness," 1 *Virtual Reality World,* page p, at q, (Summer 1993).

[37] Dorsett, Robert D., "Safety in the Air," 37 *Communications of the ACM* 146 (Feb. 1994).

[38] Karnow, Curtis, "Technology Rights in the International Arena," in this volume.

[39] Fisher, Lawrence M., "Using 'Usability' To Sell Spreadsheets to the Masses," *The New York Times* 12 (Feb. 6, 1994).

[40] Karnow, Curtis, "Alters," in this volume.

[41] *Business Week* (Jan. 24, 1994).

[42] *Business Week* (Jan. 17, 1994).

[43] Louderback, Jim, "Time To Get Smart About Controlling Your Agent," *PC Week* 92 (Oct. 11, 1993).

[44] Waldern, Jonathan D., et al., "A Note on Software Design of Virtual Team-mates and Virtual Opponents," *Proceedings, London Virtual Reality Expo 94,* London: Meckler, Feb. 1994.

[45] 2.02 *Wired* 31 (Feb. 1994).

[46] Moody, Jim, "Online Scam: Danger in Cyberspace," *Multimedia World* 68, 71 (March 1994).

[47] Reidenberg, Joel, "Rules of the Road for Global Electronic Highways: Merging the Trade and Technical Paradigms," 6 *Harv. J. Law & Technology* 241 (1993).

[48] Hayes, Brian, "The Electronic Palimpsest," *The Sciences* 10 (Sept./Oct. 1993).

[49] Schireson, Max, "Decoding the Complexities of Cryptography," 11 *PC Week* 84 (Jan. 10, 1994).
[50] Barlow, John Perry, "A Plain Text on Crypto Policy," 36 *Communications of the ACM* 21 (Nov. 1993).
[51] Levy, Steven, "Crypto Rebels," 1.2 *Wired* 54 (May/June 1993).
[52] Galvin, Chris, "The Digital Deadbolt," *Compuserve Magazine* 19 (Nov. 1993).
[53] Erickson, Jonathan, "Cryptography Fires Up the Feds," *Dr. Dobb's Journal* 6 (Dec. 1993).
[54] David Hume, *Enquiries,* L. A. Selbey-Bigge (ed.), Oxford: Oxford University Press, 1902, pp. 152–53, 159
[55] Hume, David, *A Treatise of Human Nature,* E. C. Mossner (ed.), Middlesex, UK: Penguin Books, 1969, pp. 306–310.
[56] Minsky, Marvin, *The Society of Mind,* New York: Simon & Schuster, 1985.
[57] Dennett, Daniel C., *Consciousness Explained,* Boston: Little, Brown, 1991.
[58] Ludwig, Mark, *The Little Black Book of Computer Viruses,* Tucson, AZ: American Eagle Publications, 1991.
[59] Singleton, Andy, "Genetic Programming With C++," 19 *Byte* 171 (Feb. 1994).
[60] Holland, John, *Adaption in Natural and Artificial Systems,* Cambridge, MA: MIT Press, 1992.
[61] *PC Week* 81 (Dec. 27, 1993).
[62] Flower, Joe, "Iridium," *Wired* 72 (Nov. 1993).
[63] Karnow, Curtis, "Implementing the First Amendment in Cyberspace," in this volume.
[64] Brunner, John, *The Shockwave Rider,* New York: Ballentine Books, 1975, p. 137.
[65] Rieff, Philip, *Fellow Teachers,* New York, 1972, p. 103.
[66] Dewey, John, *The Public and Its Problems,* Chicago, 1927, p. 186.
[67] Krogh, Christen, "The Rights Of Agents," in *Intelligent Agents Volume II— Proceedings of the 1995 Workshop on Agent Theories, Architectures, and Languages* (ATAL-95), M. Wooldridge, et al. (eds.).

Liability for Distributed Artificial Intelligences

11

[W]hat is interesting here is that the program does have the potential to arrive at very strange answers...

—Douglas Hofstadter [1]

Even the most intelligent among machines are just what they are—except, perhaps, when accidents or failures occur.... Could it be that artificial intelligence, by manifesting this viral pathology, is engaging in self-parody—and thus acceding to some sort of genuine intelligence?

—Jean Baudrillard [2]

INTRODUCTION

Developments in the branch of computer science known as *artificial intelligence* (AI) have passed beyond the boundaries of the laboratory; early samples are now in widespread commercial circulation.[1] At the same time, important new directions in AI research imply that the last decade or so of disappointing results will give way to truly creative, arguably intelligent programs.[2]

1. See, e.g., [3] on manufacturing, consumer products, finance, management, and medicine and [4] on manufacturing in the chemical and petrochemical industries.

2. This chapter does not address issues relating to consciousness, or programs' intentions; that is, whether computing systems can "think," "intend," or have purpose, or have a mind. See generally [5]. The works of Hofstadter [1] and Dennett [6] together suggest that "consciousness" is not *sui generis* and might emerge from a sufficiently complex and creative intelligence. Others vociferously disagree [7]. Occasionally, perceived disagreements in this area stem from the ambiguous use of the term "intelligence." Recently the achievements of "Deep Blue," a chess-playing program, have

Taken together, these two developments strongly suggest that in the next decade AI will be able to supply genuinely useful decision-making programs that operate in the real world and make decisions unforeseen by humans. This chapter forecasts the behavior of these intelligent programs and argues that they will inevitably cause damage or injury. The chapter goes on to suggest that, in the context of litigation stemming from such damage, insuperable difficulties are posed by the traditional tort system's reliance on the essential element of causation. Following the author's preference for technological answers to technological problems, this chapter concludes by offering not a new legal model, but a technological mechanism to compensate those injured by the unpredictable acts of artificial intelligences. This proposal, termed the "Turing Registry," is distinct from usual insurance schemes. In contrast to the traditional tort system, the Turing Registry proposal does not depend for its efficacy on establishing a proximate causal relationship between the injury and the agent of machine intelligence. This solution pretermits the issue of whether a specific intelligent program is "responsible" for the injury.

Complex Digital Systems

This chapter assumes that we will indeed employ AI throughout the economy, for our infrastructure is increasingly found in complex digital systems. Business, the professions, and personal work increasingly depend on the processing of computer data. The power of the digital machine allows an exponential increase in complexity, which in turn requires increasing computer power and in any event makes it impossible to turn back to manual processing.[3]

The notion of "complexity" is elusive, having technical meanings in a variety of disciplines.[4] As used here, "complexity" connotes multiple interacting but independent elements. For example, a society may be thought of as a combination of interacting but independent persons. The

---

spawned discussion on the meaning of intelligence in the machine context. See [8] (interviewees variously ascribed the label "intelligence" to purely brute-force approaches to solving complex problems, such as certain concrete chess problems, to accurate positional judgment in chess not a function of brute-force calculation, and to the ability to feel emotions or write music). This article uses the term "intelligence" as set out at note 29 below and at the text associated with note 45.

3. See generally [9].

4. For example, the Center for Nonlinear Studies at the Los Alamos National Laboratory in New Mexico investigates physical complexity in the behavior of fluids, gases, and granular materials such as sand as well as in mathematical representations. Contact office@cnls.lanl.gov. These technically complex systems can evidence chaos, turbulence, and emergent pattern formation. See generally [10].

sum behavior is a function of interactions with one's fellows as well as many individual characteristics. A car, and even more so an airliner, qualify as complex systems. As complexity increases it becomes difficult, and sometimes impossible, to predict the sum state of the complex system.

Complex computing environments are a function of the number of linked processing elements (both hardware and software) and persons (users and programmers). Plainly, this type of complexity is escalating. For example, the number of small controller computing chips (selling in the range of $2 through $5) is likely to increase dramatically over the next decade. In 1970, the typical car, office, and home had no such chips integrated into its systems. About 100 such chips per unit resided in cars, offices, and homes in 1990, and about 300 chips per unit are forecast for the year 2000 [11]. This is a part of a trend known as "ubiquitous computing," intensively studied by Xerox PARC in Palo Alto [12].[5]

These developments echo the spread of personal computers (PCs), which since 1981 show a total rise from zero in 1981 to about 100 million in 1989 to about 180 million in 1993.[6] The number of transistors on Intel chips has risen exponentially from about 10,500 transistors on the 8088 chip through one million transistors on the '386 chip to the projected 100,500,000 transistors of the P7 chip.[7] The number of lines of code in operating systems for personal computers has risen from less than 100,000 in early versions of DOS (about 15 years ago), to three million lines in Windows 3.1, to roughly 10 million lines of code in the current Windows 95 [17].

These somewhat arbitrary measures do not begin to capture the complexity at issue. Far more importantly, the number of domains of human endeavor taking place in the digital context has rapidly increased as well. It is now trivial to note that commerce, from advertising to banking and credit transactions, flows in a digital environment. Entertainment is both created and provided electronically, and social interactions—from war to art to intimate associations—are increasingly electronically mediated.[8]

---

5. James Gleick has written an amusing item describing the proliferation of very small computerized devices that we append to our bodies [13].

6. This total includes Apple computers, Intel-based computers (83% of the 1993 total), and others [14]. Worldwide sales of PCs may swell from $95 billion in 1994 and $116 billion in 1995 to $185 billion in 1999 [15].

7. Personal communications from Intel on file with author. The intermediate P6 chip has about 5.5 million transistors [16].

8. See generally [9,18]. See also [19] on the convergence of telephony and other systems.

The movement to networked systems enables this process.[9] Our computer systems do not stop at the boxes on our desks; they reach and meld with other systems around the world. Both data storage and data processing occur in those multiple remote locations, the congregation of which some call cyberspace. The number of online service subscribers worldwide is forecast to rise from 10 million in 1990 to about 55 million in 1998 [22]. Host computers on the Internet have been estimated to be multiplying at the rate of 9% to 12% percent *per month* [21].[10] Increasingly, corporations are accessing *distributed* data. These systems manage information residing on physically separated machines, although to the user everything appears to be local [23].

That expansion of networked power comes with a price: the magnification of complexity. We use a multiplicity of data formats, operating systems, and applications. We employ a variety of communications protocols, not to mention the infinite combinations of hardware. And we produce, maintain, modify, and feed into these linked systems a practically unlimited amount of data.[11]

The inextricable complexity of essential digital equipment can make life miserable for the humans who are expected to operate the systems. These human operators will not reject the help of imaginative and creative programs that seem to know their way around the electrosphere. Such programs, currently used to search and filter information from the Internet, have been dubbed "intelligent agents." Intelligent-agent technology "is going to be the only way to search the Internet, because no matter how much better the Internet may be organized, it can't keep pace with the growth in information" [25].[12]

---

9. See [20]. Orfali projects "exponential network growth" integrating, for every user, many hundreds of programs and distributed data from around the world. See also [21].

10. Further information is available at ftp.misc.sr.com/pub/zone

11. The now-standard networked client/server environment consists of many scores of interacting program modules, quite aside from the switching hardware, which acts as a program in its own right. Those modules may include system management applications such as configuration managers, storage managers, software distributors, license managers, and print managers; network management frameworks such as Hewlett Packard's OpenView or IBM's Netview; perhaps hundreds of client applications such as word processors, spreadsheets, drawing programs, desktop publishing programs, and communication programs; client operating systems such as Windows, OS/2 Warp; network operating systems such as Windows NT Server, Vines, and Appleshare; server operating systems that may or may not include the client operating systems; server applications such as Sybase SQL Server, Oracle 7, and dozens of others; and so on. Each of these programs contains its own program and data modules, some but not all of which are shared with other programs. See generally [24].

12. See also [26]. These agents are discussed at greater length below.

The Future of Intelligent Systems

The programs that promise to be most useful are not the classic "expert" systems[13] that mechanically apply a series of rules to well-defined fact patterns. To be sure, those expert systems are particularly useful in three contexts: first, where the system embodies years of human experience that otherwise might not get collected or analyzed;[14] second, where speed of operation is essential, such as an emergency nuclear reactor shut-down procedure;[15] and third, where it is cheaper to use unskilled labor to implement an expert's recorded knowledge than it is to hire the human expert.[16]

But far more useful will be autonomous "intelligent agents." The AI programs of interest, the successors to today's intelligent agents, will collect information without an express instruction to do so, select information from the universe of available data without direction, make calculations without being told to do so, make recommendations without being asked, and implement decisions without further authorization.[17]

A simple example of such a system is an airline reservation program.[18] The system would have access to a user's phone calls made through a computer program, the user's travel itinerary (available on a computerized calendar) and other computerized information. The system would note the correspondence between various calls to people in the 213 area code and trips planned for the following Friday. The program would call into United Airlines' computerized reservation system and book an appropriate flight and perhaps a hotel room for the user.[19] Details

---

13. See below, the discussion of classic expert systems.

14. See, e.g., [27].

15. Another example is to provide the pilot of a complex, high-performance aircraft "with enhanced situational awareness by sorting and prioritizing data, analyzing sensor and aircraft system data, distilling the data into relevant information, and managing the presentation of that information to the pilot" [28].

16. See generally [29].

17. For a more formal list of what we might expect from these programs, see [19].

18. For convenience, I suggest here a single "program." As noted later, a more accurate notion is an *ensemble of programs* acting in conjunction, constituting a processing environment. The inability of the traditional tort system to evaluate the actions of these ensembles results from the fact that they are indeed ensembles with disparate etiologies, not lone programs authored by a single person or company and residing at a single physical location.

19. See [30]. Some hotels permit Internet access to their reservations systems and are now moving toward the integration of these networks, with computerized property-management systems in charge of group sales and catering, remote check-in and check-out, credit-card authorization and settlement, food and beverage management, and database marketing [31].

of a more complex, if hypothetical, intelligent system are provided below.[20]

Future intelligent programs will not be cosmetic, entertaining flourishes on the front of dumb applications. Rather, the programs will truly execute their decisions with real data in a complex networked environment and will affect real world events. We already have the forerunners in mundane operation. The New York Stock Exchange, large passenger airliners such as the Airbus, the telephone,[21] and electric grids, and other computerized applications are all entrusted with decision-making authority affecting real money and real people. Some of these systems have already manifested unexpected behavior.[22] That behavior is considered an aberration; when it happens, steps are taken to correct the aberration and eliminate the unexpected actions of the software.

The legal system thinks it knows how to handle unpredictable systems. When mistakes are made, one simply traces back the vector of causation to the negligent human agency that caused the error. Then, in theory, one sues that agency for damages and is made whole. The sins of omission and of commission are just as subject to legal condemnation as negligence, recklessness, intentional malfeasance, or other human culpability.

However, some systems may be designed to be unpredictable. In addition, some decision-making programs will be highly distributed.[23] In these cases, results will be derived from a large number of concurrently interacting components, from a wide range of sources, machine and human, none alone able to make or manifest the "error." "Fixing" these unpredictable systems to operate predictably will eviscerate and render them useless. Programs flexible enough to be delegated our judgments must, of necessity, be expected to err.

---

20. See the example of "ALEF," an intelligent system sketched out below.

21. The confluence of voice and data transmission will lead to a greater integration of computers and telephony, eventually eviscerating the distinction between these systems. See generally [32–34]. Note also the expected merger of the television set and the computer [35]. The recent enactment of the Telecommunications Act of 1996, signed by President Clinton on February 8, 1996, will permit companies in the cable, television, and telephone industries to compete in all these related arenas, Pub. L. No. 104, 110 Stat. 56 (1996). Already television cable is being used to provide Internet access, and some commentators foresee an expansion of that technological merger [36].

22. See generally [36]; more details are available from the Internet USENET group comp.risks at http://catless.ncl.ac.uk/Risks and archived at ftp:unix.sri.com/risks. RISKS-LIST: Risks-Forum Digest is also available at this address. The unavoidable risks posed by complex software are discussed further below.

23. *Distributed* intelligence suggests a series of programs physically resident at various disparate sites, interacting with each other to present the user with a single integrated result. See, e.g., [37].

Under these circumstances, the law may hesitate to make a simple assignment of responsibility. It is not clear what the law will or should do when artificial intelligences make mistakes, thereby damaging property, causing monetary losses, or killing people. Perhaps we will blame nature or the inchoate forces of the universe. But the legal system is unlikely to rest there; we will not long accept equating the damage done by an unexpected tornado with the mistakes made by programs that are, at some level, human artifacts. If someone *might* have been able to control the outcome of a series of events, the law is likely to be invoked when that control is not exercised. Even with natural disasters, those who could have forecast the path of a storm,[24] or warned of the danger of down drafts and wake turbulence,[25] have been sued. It does not matter that the specific persons responsible cannot be identified. Many legal doctrines assign legal responsibility when it appears that someone, somehow, in some way, was the actual cause of injury.[26]

Perhaps we should look to the collection of networked systems and operators in the midst of which intelligent programs do their work, for it is in this ambiance that the intelligent program operates. No one system, and no one systems operator, programmer, or user, will know the full context of a networked intelligent program—managing that complexity is precisely why the program was employed. Yet where responsibility and thus liability are so spread out over numerous and distributed actors, the very sense of "responsibility" and "liability" is diminished.[27] At that point legal causation seems to fade into the background and tort law, founded on the element of causation, falls away.

Assigning legal liability involves discrimination among an infinite number of causal candidates. That discrimination is avowedly based on perceptions of policy—society's collective sense of what is reasonable and who should be blamed for certain injuries. This chapter suggests that advances in AI, specifically in the distributed computing environment in which such programs will operate, will eviscerate the very idea of cause and effect. Where there are no grounds on which to discriminate between "reasonable" and "unreasonable" candidates for blame, the notion of

24. *Brown v. United States*, 790 F.2d 199 (1st Cir. 1986) (holding defendants not liable for drowning of fisherman in a storm that the National Weather Service did not predict).

25. See *Wenninger v. United States*, 234 F. Supp. 499 (D. Del. 1964).

26. See, e.g., *Summers v. Tice*, 199 P.2d 1 (Cal. 1948).

27. We are already familiar with an incipient version of this. Users of home PCs are routinely told that their problems have to do with the complex interaction of CD-ROMs, operating systems, memory management software, IRQ and other settings within the machine, various peripherals, sound cards, graphical user interfaces, and other applications, and on and on.

"legal cause" will fail, and with it the tort system's ability to adjudicate cases involving artificial intelligences.

The legal issues discussed here cannot be solved through new legal tests, criteria, or models much beloved of law review articles. The solution sketched out in Part V is more a technological than a legal solution. That is as it should be. Good solutions to problems of advancing technology are those that do not need repeated access to the courts.

THE DEVELOPING TECHNOLOGY

Looking Back: The Classic Expert System

At some level, all computer programs make decisions of which the human user is ignorant. These are decisions apparently "made on its own" with some apparent degree of autonomy. The elemental decision tree, if/then statements, and branchings in program flow are common to most programming languages.[28] These decisions are based the current state of data, which will often be unknown to the user. The temporary state of a variable, data from an input device (e.g., a signal from a remote modem asking for a certain baud rate), and so on, are hidden from the human operator. To those ignorant of the internal workings of the program, it may seem like a "black box," a secret process that magically generates a sensible, context-accurate and apparently intelligent response.

This sort of automated response secured from our day-to-day use of computers now seems ordinary. The program's low-level decision making has in fact simply been built in by the original programmer. Mundane programs, such as word processors, spell checkers, accounting programs, and even grammar advisors, are at the lower end of a spectrum of "self-directedness" or automation. These programs produce unexpected results only in trivial ways, such as when our fingers slip as we type. They may be quicker at their limited tasks than humans, and they do not suffer the vice of indolence. But they act only on direct command and use only spoonfed information.

Next along the spectrum of self-directedness are so-called "expert systems." Various technologies are used to achieve an expert system, some of which can very roughly be termed intelligent. The honorific is used because these systems attempt "to generate heuristics, or rules of

---

28. Users might be aware of, and indeed command, higher order decisions such as "*if* the temperature is above 43 degrees *then* close the circuit." But virtually all program decisions are of a lower order, such as "*if* certain address in memory contains the datum XX *then* replace it with datum ZZ," or "*if* a datum at memory location is XX *then* jump to a different section of this program and execute the instruction found there."

thumb, to guide the search for solutions to problems of control, recognition, and object manipulation" [38]. In short, we term "intelligent" those systems that appear to mimic the higher cognitive abilities of humans.

A wide variety of programming techniques may be applied to the goal of making an artificial intelligence.[29] One now-classic technique is a neural network. Trained neural networks contain so-called hidden layers of weighted nodes, which interact to generate certain output under various conditions. During a training period of constant feedback from the human "trainers," these neural nets experiment with various values to their internal nodes, until the net combination of these values generates, in enough cases, the result the trainers desire. The weights on values are then fixed, new imputs are provided, and the trainer then expects results comparable to those secured in the training sessions. [30] "The problem of learning in neural networks is simply the problem of finding a set of connection strengths [in the nodes] which allow the network to carry out the desired computation" [41]. The specific configurations of the nodes and their weights are irrelevant and usually not known to the operator.[31]

---

29. The wonderful and nebulous term *artificial intelligence* covers a plethora of programming techniques and goals. Direct modeling of the activity of the brain's network of neurons is one such field; also included are *case-based* reasoning systems, which try to apply rules to new factual scenarios. AI systems are used to model complex movement in robotics, including the difficult areas of perception and visual and aural discrimination [39]. See also [4] (describing hybrid AI systems using a combination of neural nets and other programming techniques to control complex manufacturing processes). As this chapter notes later, the term "intelligent" in the context of so-called intelligent agents refers to agents (program chunks or *modules*) that can communicate and work with other agents to produce a larger program, which in turn appears to mimic the higher cognitive abilities of humans.

30. Here is how one team of researchers introduced the idea of neural nets [40]: "[T]he most common models take the neuron as the basic processing unit. Each such processing unit is characterized by an activity level (representing the state of polarization of a neuron), an output value (representing the firing rate of the neuron), a set of input connections, (representing synapses on the cell and its dendrite), a bias value (representing an internal resting level of the neuron) and a set of output connections (representing a neuron's axonal projections).... Thus, each connection has an associated weight (synaptic strength) which determines the effect of the incoming input on the activation level of the unit. The weights may be positive (excitatory) or negative (inhibitory). Frequently, the input lines are assumed to sum linearly yielding an activation value."

31. The physical appearance of a neural net need be little different from that of any computer. It may have a camera, for example, to enable the input of digitized video. The nodes are simply values (if that is not too loose a term) embodied in software. For a discussion of neural nets generally, see the Frequently Asked Questions (FAQ), available at ftp:rtfm.mit.edu/pub/usenet/news.answers (filename neural-net-faq). See also http://wwwipd.ira.uka.de/~prechelt/FAQ/neural-net-faq.html on the World Wide Web. There is a good deal of literature on the subject of neural nets. See generally [19]

For example, one may show a neural net system a series of pictures of faces and tell it which faces are of the same person, or the net may examine a series of pictures of healthy and diseased organs. In each case, the net is told the correct answer during training, and the net then internally adjusts itself to correlate the input with the correct output. Subsequently, the net will take new input, new pictures of faces or of diseased organs, or perhaps numbers from the stock market, and then generate new correlating output as a function of its "learned" internal states. Thus, the neural net might conclude that a face was indeed of a certain notorious criminal, that an organ was cancerous, or that a stock should be bought or sold. "A neural network can perform highly complex mappings on noisy and/or nonlinear data, thereby inferring subtle relationships between sets of input and output parameters. It can in addition generalize from a limited quantity of training data to overall trends in functional relationships" [45]. Thus, for example, neural nets have acted as expert mortgage insurance underwriters, vehicle operators, and sonar processing systems [46].

These neural net systems work as long as the type of input and the general context for the expert system closely parallel the type of input and context of the training sessions. The systems are optimized for specific tasks: for example, playing chess, providing medical diagnoses, or adjusting the control surfaces of an airliner. These systems do not function at all in a different context. They do not cause unexpected results except in failure mode. Even though the machine's operators do not know the contents of the "black box," they do know the program's purpose and limits. Obviously, it is wrong to use a car engine diagnostic system to interpret a person's medical condition; it is wrong to use a system expert in the game of *Go* to make judgments about stock market investments. If loss of life or money result from that "misuse" of a computer system, we know to blame and sue the operator.

These expert systems (whether neural nets or not) are truly machines in the old-fashioned sense. They are housed in specific hunks of metal and silicon, and fed carefully culled information in meticulously and specifically prepared chunks. These exhibit "intelligence" in only a weak way, regurgitating intelligence like a child imitating an adult. This lack of meaningful intelligence is patent when programs randomly combine words and then edit those according to rules of programmed grammar to generate "poetry." The same lack of meaningful intelligence

and [42]. Neural nets can be used to analyze highly complex groups of constraints: for example, motion controls for robots that require the analysis of feedback from the environment present difficult nonlinear control issues, solvable by such nets [43]. See also [44] (describing how the many factors that go into the calculation of efficient chip fabrication can be solved with these expert nets).

is evident in programs which, given the rules of logic, spit out syllogisms, or make a correlation between objects whose common properties were already programmed.

These sorts of programs, predictable, specialized in task, able to use only specially prepared data, and only weakly intelligent, are not very interesting in the present context.[32] But, as a result of just those qualities, the programs present no serious conceptual difficulties for the legal system when people lose their lives or property as a result of their decisions. That is, when damage results from the employment of a neural net, it is not difficult to trace back causal vectors to the program, to its trainers/programmers or to its users.

Looking Forward: Fluid Systems

Recent work, best exemplified by Douglas Hofstadter and his group at the University of Indiana [1], points to the development of creative, intelligent programs designed to handle radical shifts in context, and to produce useful and creative output. They are thus designed to produce unpredictable but pertinent results. These programs, and their expected progeny, deserve to be called "creative" and perhaps "intelligent" in a strong sense.

Hofstadter describes intelligence as emerging from thousands of parallel processes that take place in milliseconds. These processes are generally inaccessible to introspection [47]. Modeled by computers, such intelligence [48] is not directly programmed or "hard-wired" (in contrast to many of the "expert" systems referred to earlier), but emerges as a statistical consequence of the way in which many small program fragments interact with each other. This leads to what Hofstadter terms "epiphenomenal" or emergent intelligence. His programs excel (albeit on a small scale) in *making analogies*; that is, in being able to examine a wide variety of input and relating it in different ways depending on the context, extracting and utilizing different properties of the data from time to time. The context or universe reciprocally derives from the programs' initially tentative, mutable examination of the data.

Hofstadter notes that in the context of human perception, our assumptions are modified by what we see. What we discern is dependent

---

32. Such systems include what professor Jim Bezdek terms a basic "computational neural network." These give the appearance of intelligence, but they operate "solely on numerical data such as that obtained from sensors and computational pattern recognition ...produc[ing] numerical results that reveal structure in sensor data" [46]. The ability to manage substantial combinatorial complexity, though, is not the exhibition of intelligence or true learning. "This idea that neural networks learn is just stupid," Bezdek notes. "They don't learn.... An algorithm is just an algorithm" [46].

on that context and on the ability to make analogies to generally persisting analytical structures. For example, as we are increasingly exposed to music, we are able to discern notes and tonalities to which we were previously "tone deaf." Experienced pilots often have unarticulated assumptions about what a cockpit environment should look and feel like, created by sensory input during the course of their flight training. Conversely, a pilot's ability to discern changes in his environment, a musician's ability to discern tonality shifts, and a doctor's ability to read an x-ray picture are all functions of preexisting knowledge structures. These structures persist until modified by overpowering new sensory input.

Where humans lack the appropriate preexisting analytical structure, the data is not recognized as "relevant." In short, perception is a function of making analogies.

> The ideal self-organizing network should be able to use context and historical experience to decide what it will learn and ignore. In other words, it should be able to determine for itself what inputs are significant and what are not. In our everyday experience, whether we label a particular piece of sensory information as meaningful or irrelevant often depends on the context [49].

"[I]n a complex world...one never knows in advance what concepts may turn out to be relevant in a given situation" [50]. Intelligent programs must be able to look at data from a variety of perspectives and extract and analyze a variety of properties from the data depending on the nature of the problem to be solved.

The entire system's activity is simultaneously top down and bottom up. That is, the input selected as "relevant" from the entire universe of potential data is influenced by preexisting but not ultimately permanent structures or assumptions, and those structures are simultaneously modified by new input as "learning" takes place [51].[33] Contexts used to order new data can be dislodged and replaced under the influence of new input. This becomes increasingly difficult, however, as an existing context gets reinforced through the processing cycle and patterns in data are discerned under the framework of that context. Thus, contexts, concepts,

---

33. Professor Hofstadter in [51] describes parallel architecture by which bottom-up and top-down processing influence each other. See [52] (describing a computer neural network by which bottom-up "recognition" connections convert input vectors to representations in one or more hidden layers of the neural net. The top-down "generative" connections are then used to reconstruct an approximation of the input vectors from the underlying representations).

and perceived data are constantly cycling, slipping past each other, and occasionally agglutinating and falling apart, until a "probable best fit"—the most unifying answer—is found or another terminating event occurs. Hofstadter's model of innumerable hidden or unconscious subprograms generating many stochastic microdecisions without a supervising or top-level executive director corresponds well with other models of mental activity [53,54].[34]

But this "probabilistic halo" [55] model of truly creative intelligence has more important consequences. While the basic statistics do not change, some bizarre and improbable "fringe" responses very occasionally appear on repeated runs of the program [56]. Hofstadter notes that "[i]t is critical that the program (as well as people) be allowed the potential to follow risky (and perhaps crazy) pathways, in order for it to have the flexibility to follow insightful pathways" [56]. The desired strongly emergent behavior results when "one or more aspects of [the system's] behavior is theoretically incalculable" [57]. Specifically, the system may select an inappropriate context and thus may neglect data because the data did not fit a context or analogy that fit other data. Alternatively, the system may reject a good analogy because of peculiar or aberrant data.

It is an essential and, over the long run, entirely assured characteristic of intelligent programs that they will make "pathological" decisions. The nature and timing of these pathological decisions cannot be known in advance. These are not "bugs" in the programs, but are part of their essence. True creativity and autonomy[35] require that the program truly make its own decisions, outside the bounds expressly contemplated by either the human designers or users. Hofstadter's programs do just that, and I suggest they are the precursors of tomorrow's commercially available AI programs.

The Unreliability of Software

The failure of a complex program is not always due to human negligence in the creation or operation of the program, although examples of such negligence are legion.[36] But, in addition, there are *inherent* problems with

34. See generally [6].

35. "Autonomy" is a relative term. Some systems are more or less autonomous than others, and there is a broad range of autonomy and automation [58]. I have not expressly treated that range in this chapter. I have addressed the issue below instead through the prism of multiple concurrent causation, which recognizes variable contributions of humans and machines to a given result.

36. Buggy software is released even when known to be buggy. Corel was told of crashes in its drawing program, but released it anyway, apparently to meet market deadlines [59].

software reliability.[37] While it is at least theoretically possible to *check* to see if a program output is correct in a given instance, it has not been proven that programs can be verified as a general matter; that is, that they are correct over an arbitrary set of inputs. In fact, it appears highly unlikely that even programs that successfully process selected inputs can be shown to be correct generally [61].

Software reliability generally cannot be conclusively established because [62]

> digital systems in general implement discontinuous input-to-input mappings that are intractable by simple mathematical modeling. This...is particularly important: continuity assumptions can't be used in validating software, and failures are caused by the occurrence of specific, nonobvious combinations of events, rather than from excessive levels of some identifiable stress factor.

The long-term operation of complex systems entails a fundamental uncertainty, especially in the context of complex environments, including new or unpredictable environments [63]. That, of course, is precisely the situation in which intelligent agents are forecast to operate.

An excellent overview of the difficulties in checking a program has been provided by Lauren Wiener [64]. It is, she notes, practically impossible to test software thoroughly. To test a program, all possible sequences of instruction must be run to see if the program in fact behaves as it should. This can take literally thousands of years [65]. For example, assume a stunningly simple program with (1) between 1 and 20 commands, (2) that may be called in any order, and (3) that may be repeated any number of times. For a thread (or sequence) of execution one command long, there are of course exactly 20 possible threads. For a thread two commands long, we have 20 × 20 or 400 possible threads. Those familiar with the mathematics of combinatorial explosions will see this coming. As the number of commands in the thread goes up, the number of threads that need to be run and tested rises exponentially.[38] Spending one second per test run, it would take 300,000 years to test this simple program of between 1 and 20 commands. If a bug is found, the entire testing cycle may have to be repeated since a bug fix means, by definition, a program change.

37. See generally [60].

38. The number rises from 20 (one command) to 400 (two commands) to 8,000 (three commands), and finally to 10,240,000,000,000 (ten commands) [65].

The problem is not limited to what is conventionally thought of as software. The hardware on which these programs run, the processing chips themselves, can be thought of as reified programs, programs encased in silicon. And despite extensive testing by their manufacturers, those chips, too, are buggy [66].[39] It is thus no surprise that, from time to time, software fails, property is damaged, and people are killed.[40] Consequently, programmers just do the best they can. They try to implement rules of thumb and common sense, keeping programs as simple as possible, documenting the code, creating initially stable designs and so on [68].

We see that it is practically impossible to debug a fixed program with a known range of inputs, on a fixed, unblemished platform. This chapter, however, posits the interaction of multiple "intelligent" programs on unknown platforms where none of the programs, operating systems, and chip architectures is known in advance, where each agent may provide nonpredicted data for input to the other agents which are at that point part of the ad hoc ensemble. It is fair to suggest, then, that although we are assured that intelligent agents will malfunction, we cannot possibly be expected to foresee the nature or timing of the particular problem.

Multiagent Networked Collaboration

Much has been written on so-called intelligent agents.[41] These agents are programs originating at one site and executing at a different site. A current model comprises a program written at a local computer, translated into data packets, and sent into the telecommunications network, then reassembled at the target computer into a program whose instructions

39. This reference notes errors in the initial batch of Intel's Pentium Pro (known in development as the P6) which could corrupt data, as well as recalling problems with the original Pentium chip. The first releases of the Pentium chip (originally known as the P5) in late 1994 contained a floating point unit (FPU), or arithmetic coprocessor, which was comprised of registers unable to correctly handle a divide function. This was the so-called FDIV error. See the report of the Pentium bug [67].

40. The Federal Aviation Administration's Advanced Automation System's millions of lines of code are buggy, and so the nation's skies remain under the control of decades-old computer technology. When three lines of code were changed in a telecommunications program in the summer of 1991, California and much of the Eastern seaboard lost phone service. Two cancer patients were killed in 1986 by overdoses of radiation from a computer-controlled radiation therapy machine; errors included a so-called "race" condition (where two or more threads in a parallel processing system do not execute as predicted) [68]. This race condition is described in Part IV of this book.

41. See generally [69–72], along with http://www.doc.mmu.ac.uk/STAFF/mike/atal95.html and http://www.cs.umbc.edu/agents on the World Wide Web. A recent study suggests (perhaps with a bit of exaggeration) that "[a]gents will be the most important computing paradigm in the next ten years" [73].

are executed at the target system.[42] These programs can secure information and then act on it, such as locating flight information and making reservations.[43]

This point is worth reemphasizing: agents are programs, originating at one site and executing at a different site. Agents are cross-platform compatible; host machines can run agents from other sites regardless of the otherwise incompatible hardware and operating system. Users will often not know when or where their agents are executing [79].

Agents interact with other agents. This is analogous to the way in which mundane programs, such as word processors and telecommunications programs, are comprised of a large number of relatively autonomous subroutines that interact with a variety of other programs, such as macros, spell checkers, file managers, and file transfer programs. Agent technology expedites the sharing of large tasks and information and multiagent collaboration.[44] Agents that "know" how to identify and work with other agents—so-called "intelligent" agents—will be more useful and popular than those that do not have such abilities. This ability to collaborate with other agents toward a larger goal may be termed "intelligence" [83,84]. "A group of agents or processes is almost always more successful at solving a problem than single agents or processes working in isolation" [85].[45]

By design, an agent's creator or user need not know where the agent goes to do its work or know the other systems and agents with which it will interact. This recalls aspects of today's Internet environment, where users do not know which computers are sending them files or which machines are receiving their mail. Users are generally ignorant of (and indif-

---

42. See [74].

43. Hot Off The Tree (HOTT), April 25, 1994 (Internet download), carried a report describing an information-gathering program known as Hoover, from Sandpoint Corporation (Cambridge, MA). Hoover's search results are compiled into a customized electronic newsletter, with headlines that can be clicked on with a mouse to retrieve full-text articles. Microsoft's Office suite includes Intelligence for real-time spelling error correction. Other software packages include Beyondmail from Beyond, Inc. and Open Sesame! from Charles River Analytic (Cambridge, MA). Beyondmail automates responses to incoming email. Open Sesame! monitors repetitive PC activity and then, in essence, automatically creates intelligent, autonomous macros. AT&T planned to introduce Telescript email in 1994, which would have allowed users to type in the addressee's phone number; an agent would then look up the email address corresponding to that number and deliver the message to the addressee's computer. General Magic plans to license Telescript freely to other companies. Presumably the system will allow cross-platform, network-independent messaging, insulating users and programmers from the complexities of network protocols [75–78].

44. See [80–82].

45. See also [86]. I am grateful to Mr. Huberman, who is a research fellow with Xerox PARC in Palo Alto, California, for providing me with his articles.

ferent to) the multiple programs executing to serve their queries and route their messages.

Current research points directly toward the use of these distributed agents, the *ensemble* of which will achieve the general goals desired by human users. These components will "intelligently" cooperate with each other, securing relevant data from each other, often interacting with each other "in unanticipated ways" [87] across platforms and operating systems. "We envision a world filled with millions of knowledge agents, advisers, and assistants. The integration between human knowledge agents and machine agents will be seemless [sic], often making it difficult to know which is which" [88].

Current work at Sun Microsystems on the new programming language Java illustrates (and will enable) the movement toward multiagent collaboration across networks.[46] Java is derived from the relatively familiar C and C++ programming languages. Java programs spawn small programs or "applets" to be downloaded across the Internet, creating distributed dynamic programmed content.[47] Java does this without regard to the platform or operating system on the remote sites (i.e., Java is system or platform independent) [92]. An obvious application is the creation of intelligent agents [93]. There are other projects for generating distributed objects across the Internet as well [94], although with Microsoft's endorsement Java will now probably become the standard [95]. Users have recently discovered that Java applets may pose serious security problems as they interact with Internet browser software in unanticipated ways.[48]

Agents and Networked Viruses

As discussed earlier, agents are programs created at one site but designed to operate at other sites. Thus, some have noted that an agent is much like

46. See generally [89,90].

47. These distributable agglutinations of data and program code are known as *objects*. The notion of combined data and programming code is embodied in another term used by Java programmers, "executable content" [91].

48. Java is fast and reportedly highly secure. Recognizing the risks posed by allowing applets—true blue executable programs—to download from remote sites into users' host machines, one title blandly assures the reader that "No Java applet is able to steal information or damage your computer in any way" [96]. Security is achieved by (1) restricting the environment in which Java applets can run (a notion treated under the rubric of "global controls" later in this chapter) and (2) verifying the individually transmitted "bytecodes" that are subsequently interpreted and then run together on the user's host machine (approximating the certification process discussed in this chapter) [97,98]. However, more recently, security flaws have been discovered as Java interacts with Netscape's browser software [99]. See also [100]. Details and further discussion may be found in RISKS-LIST: Risks-Forum Digest, vols. 17.83 and 17.85, available from the Internet USENET group comp.risks at http://catless.ncl.ac.uk/Risks

a virus [101]. There is a very thin—some would say imperceptible—line between viruses and presumably "beneficial" programs. "The biggest danger of any network-wide system that allows intelligent agents is that some of the agents will deliberately or accidentally run amuck. Agents have much in common with viruses: Both are little programs that get to seize control of a foreign machine" [102].[49]

As with "flowers" and "weeds," the definition depends on what one desires. A recent report expressly conflates the two notions [104]:[50]

> Now, Japan reports the first "beneficial virus." A report in the *Nikkei Weekly* claims that a group from the Tokyo Institute has developed a virus that travels through computer networks, collects information, spots glitches and reports its findings back to network managers. Wonder if it fetches passwords too.

Many commercial programs have undesirable virus-like side effects, known by some but not all users: networks can crash, files can be deleted, and data can be mutated.[51]

The problem, as always, is that the circumstances of an agent's remote execution, including the other agents with which it may come into contact, cannot be fully predicted by the agent's owner or the remote system's owner. An excellent example was recently provided by the appearance of the *macro* virus. Programs such as Microsoft Word have a built-in scripting (programming) language that allows the creation of small mini programs known as macros. Other popular programs such as Borland's Quattro Pro spreadsheet and WordPerfect have similar macro capabilities. These macros simply allow the user to do with one command what might otherwise take a repetitive series of commands. In Word, however, these

49. Wayner does explain that the version of agent software he examines (TeleScript) has a "security" system by which the host computer can allow only previously authorized users to send in agents. Such an authentication system, however, which is also used by Microsoft's ActiveX agents, is not foolproof. Authentications can be secured under false pretences, and even a well-meaning authenticator may place his imprimatur upon a dangerous agent. See [103]. See also Microsoft's statement at http://www.microsoft.com/security

50. Neither my use of the term "virus" nor that of *InfoSecurity* is precise here. Normally machine viruses are considered to be *self-replicating*, and not necessarily unintentionally (or indeed intentionally) destructive. See generally [105,106]. However, the classic definition and my use share the connotation of a program that is (1) hidden within (and dependent upon) a software host, and (2) independent, mobile, and transfers itself from host to host.

51. See generally "Recombinant Culture: Crime in the Digital Network," Chapter 12 of this volume. See also [107]. Louderback writes that TeleScript is not a virus, "but it makes it easier for rogue programs to spread" [107].

macros can be hidden inside ordinary data files and then transmitted electronically to other computers, where the macro program is executed as the file is read.[52] An executed macro virus may change screen colors, reproduce itself, add or garble text, or cause other problems. Macro viruses, according to Karen Black of Symantec Corp., "have the potential to be a much bigger threat than conventional viruses, because people exchange data files all the time.... With the Internet, the problem becomes much worse" [108].[53]

The point of equating some general programs with viruses is to note that in a complex computing environment, the best-intentioned program may contain modules with unintended consequences. Because the true scope of the relevant computing environment includes an entire net- work or the Internet (all connected machines), the actions of multiple interacting intelligent agents must be extrapolated in a wide variety of environments.

If our most useful agents—those we send to connected machines to do the most creative work—have undesired consequences operating alone, then the problem will be exacerbated in the context of an express *community* of distributed programs interacting on an ad hoc basis: "We recently constructed a theory of distributed computation.... This theory predicted that if programs were written to choose among many procedures for accessing resources without global controls, the system would evolve in ways independent of the creator's intent. Thus, even in simple cases where one would expect smooth, optimal behavior, imperfect knowledge in cooperative systems with finite resources could lead to chaotic situations" [111].[54]

Microsoft may have provided us with an interesting precursor of the inadvertent harm that could be caused (or allowed) by an agent—in this case, one that is relatively dimwitted. Users of the Windows 95 operating system found out in the spring of 1995 that, unbeknownst to them, their

---

52. See [108,109].

53. The macro viruses appear to be a reincarnation of a problem that cropped up a few years ago in an early version of Microsoft's object linking and embedding (OLE) technology. Certain applications such as Microsoft's Office suite of programs allowed users to create, in effect, self-executing data files which would be held inside another data file (such as a letter or a chapter of a book). For example, a document might have a graphic "object" (say a picture of an apple) embedded as an illustration in the document; OLE would *execute* the object, automatically running a new program to display it, as the reader read the document. But OLE "may inadvertently create a backdoor through which malicious individuals can enter to embed destructive commands, viruses and worms in any number of applications that are distributed via electronic mail or network protocols" [110].

54. The use of what Huberman calls "global controls" would reduce the autonomy of the system and undermine its claim to collaborative, emergent intelligence. This is discussed below. See text accompanying note 112.

operating system contained an agent that was silently reporting back to Microsoft [112]:[55]

> Microsoft officials confirm that beta versions of Windows 95 include a small viral routine called Registration Wizard. It interrogates every system on a network gathering intelligence on what software is being run on which machine. It then creates a complete listing of both Microsoft's and competitors' products by machine, which it reports to Microsoft when customers sign up [electronically] for Microsoft's Network Services, due for launch later this year [1995].

Microsoft disputes evil intent and specifically denied that the Wizard can report on the hardware configuration of every PC hooked up to a network. Microsoft describes this as just an automated version of a registration card. But the Wizard does appear to detect, and report back to Microsoft, all hardware and software of the local PC.[56] The user of the program is not aware of the scope of the report; rather, the Wizard operates independently of the user for ulterior purposes.

An anonymous Internet commentator reflects:

> A friend of mine got hold of the beta test CD of Win95, and set up a packet sniffer between his serial port and the modem. When you try out the free demo time on The Microsoft Network, it transmits your entire directory structure in background.
>
> This means that they have a list of every directory (and, potentially every file) on your machine. It would not be difficult to have something like a File Request from your system to theirs, without you knowing about it. This way they could get a hold of any juicy routines you've written yourself and claim them as their own if you don't have them copyrighted.

The juvenile intelligent agent described here may: (1) perform unauthorized transmission of trade secrets;[57] (2) violate federal copyright law;[58] (3) possibly interfere with privacy rights;[59] and (4) perform an un-

55. As cited in RISKS-LIST: Risks-Forum Digest vol. 17.13, available from the Internet USENET group comp.risks at http://catless.ncl.ac.uk/Risks. See also [113].
56. Microsoft's position is available at ftp.microsoft.com/peropsys/ win_news/regwiz.txt
57. See, e.g., Cal. Civ. Code § 3426 (West Supp. 1996) (Uniform Trade Secrets Act).
58. 17 U.S.C. §§ 501-10 (1977).
59. See generally [114] (on the right to control personal information).

authorized access or interception in violation of the Electronics Communications Privacy Act and related state penal codes.[60]

Polymorphism and the Units of Programming Action

The forecast of intelligent agents provided earlier suggests that agents will constantly mutate to accomplish their purposes and might even become unrecognizable to their owners. This situation is aggravated by the agents' viral aspects noted earlier. Some viruses, both biological and digital, mutate to survive by defeating immunological and other weapons designed to destroy the viruses; it is obviously more difficult to defeat an enemy that constantly changes its guise. The assumption that intelligent programs will mutate to accomplish their purpose is strengthened by references, conscious or not, to recent work in the artificial life context. For example, interesting recent developments suggest the self-modification of code and the creation of program elements which, left to their own devices and evolution, may result in exceedingly efficient programs that humans alone could never have created.[61]

The intelligent agents forecast earlier need not be able to change or adapt in quite that manner to accomplish their purposes. However, this discussion does suggest a larger problem of mutation or *polymorphism* in programs. Polymorphism can be confusing because in the digital context it is an ambiguous notion. Specific bits of code (we might call them *codelets*, as Hofstadter does) may or may not change as a larger chunk of program mutates. The larger the program, the more likely it is that as processing takes place "the program" can be said to change in some way. Many "programs" are in fact composed of dozens of discrete codelets and data sources such as dynamic link libraries (DLLs),[62] initialization files,

---

60. 18 U.S.C.A. §§ 2510-22 (West 1970 & Supp. 1995). See Cal. Penal Code § 502 (West Supp. 1996) (discussing unauthorized access to computer data and systems). For more information on so-called web-wandering "robots," "spiders," and other programs that among other things take data from remote locations without permission, see [115], (discussing the dangers of software "robots" accessing information without human supervision and judgment); see generally [116] (citing http://web.nexor.co.uk/mak/doc/robots/robots.html on the World Wide Web).

61. See generally [117]. Professor Holland and others have experimented with systems that compel competition between programs to evaluate their suitability for certain tasks. They then use selected parts of successful programs for "genetic" recombination, creating succeeding generations of programs. These new programs compete and begin the loop again. After many iterations, highly successful programs can be generated [118].

62. See note 66 below.

and device drivers,[63] with logical program flow depending on all of these subsidiary structures. These structures, in turn, can be shared among a number of encompassing "programs."[64]

While we use the term "program" for convenience, it is more accurate to think of a *processing environment* containing both data and instructions (or a *logical program flow* depending upon both data and instructions), because no fundamental distinction can be drawn between the two. Computers see data and instructions in the same binary way. Data can act as instructions and vice versa. A program can accomplish the same result by modifying data and feeding them to a static program, or by modifying a program and feeding it static data.

Today, object-oriented programming (OOP) is common; for example, Borland's and Microsoft's C++ programming language provides OOP. This language makes express the community of data and instructions. A so-called object is a bundling of "data structure and the methods for controlling the object's data" [122]. These objects will, at least in C++, always contain both the data and the data's behavior rules. These two are *encapsulated* in the object. As an elemental programming unit, this object can then be duplicated, modified, and used in an infinite number of contexts. These objects, which are types of programs or codelets, may or may not change independently of the larger program.[65] Often the properties of the object—that is, its combined data and instructions—are invisible to the programmer and even more so to the program's ultimate user [123].

A putatively static program may operate in a dynamic (data) environment. In such a case, referring to the processing environment at large, we could as correctly classify the program as dynamic or polymorphic. When the context of a constantly mutating data environment is included in the conception of the processing environment, it becomes apparent that *all* programs should be thought of as polymorphic.

There is, to be sure, a forceful convenience in naming programs and assuming their identity and persistence through time. Without such as-

---

63. These drivers are small programs "designed to handle a particular peripheral device such as a magnetic disk or tape unit" [120].

64. "The most important event in the history of software happened somewhere around 1959, when the designers of a programming language called 'Algol 60' realized that *you can build a large program out of smaller programs*" [121]. See note 11 (listing multiple modules used in current client-server environment).

65. A similar effect can be simulated within the Microsoft operating system Windows. An "object" such as a spreadsheet or graphic can be created and then inserted into data created by another program, such as into a letter created by a word processor. This so-called "object" acts as a combination of the data (i.e., the spread-sheet values, the numbers, or the picture) *and* the program required to modify the data (the spreadsheet program or the paint program). See note 53 (OLE corruption problems).

sumptions of persistence and identity, copyright and other legal doctrines affecting software would be difficult to conceive. Most of those doctrines, after all, are based on traditional property-based notions, and they depend for their sense on persisting, identifiable, and preferably tangible property [124]. But those assumptions are just convenient fictions. As with other legal fictions of persisting entities (such as corporations), the fictions succeed to some extent, and for some purposes.

However, to the extent that the criterion of continuity and the true behavior of software agents diverge, the utility of the fiction of persistence dissolves. Even today many problems resulting in data loss and interruptions in computer services are clearly caused by ever-shifting interactions among the subsidiary "programs" or data files referred to earlier, such as initialization files and DLLs, which operate independently.[66]

In brief, processing environments are polymorphic and distributed. The environments change over time and overlap in time and in space (computer memory). We will see[67] that these characteristics pose great difficulties for a traditional legal analysis of the effects of such processing environments.

Summary Forecast

Before we move to the legal domain, we should summarize the forecasts.

We should assume that:

1. We will employ intelligent agents specifically for their creative intelligence and corresponding judgment;
2. These agents will make decisions as the consequence of the interactions of a distributed ensemble of such agents;
3. This ensemble will be comprised of polymorphic programs;
4. Some of the resultant decisions will be "pathological"; that is, not only unpredictable and surprising, which may be desirable,

---

66. For example, companies such as CompuServe and Netscape, as well as Windows' manufacturer Microsoft, sell Windows-compatible Internet communications programs. Their programs call on a library of functions—a set of small programs or modules—found in (among other places) a DLL drafted to a public technical standard known as the Windows Sockets. This library of modules is called WINSOCK.DLL. Multiple programs can call on the functions contained in WINSOCK.DLL. But when the user installs Microsoft's Internet package, the installation quietly renames WINSOCK.DLL to WINSOCK.OLD, and then copies its *own* WINSOCK.DLL to the user's drive. Programs other than Microsoft's can not use this new DLL, and so they fail [119]. As of fall 1996, at least, Microsoft had not told its competitors how to use Microsoft's superseding WINSOCK.DLL [119].

67. See note 146 below.

but also having unintended and destructive consequences both for networks and for the substantial portions of our infrastructure connected to those networks.

The question of liability for these unintended but inevitable injuries is far from simple, and cannot be solved under current legal doctrine. As discussed in the next section, "causation" analysis of these injuries is particularly difficult.

CAUSATION IN THE LEGAL DOMAIN

Men are not angered by mere misfortune but by misfortune conceived as injury.

—C. S. Lewis [131]

Where injury is done, we suppose the law should intervene. The legal system fixes liability on those responsible for the injury, the so-called "legal cause" of the injury. The drunk driver is the cause of the car accident and should pay the victim he hits. Perhaps the aircraft manufacturer is liable if the aircraft is negligently designed and crashes. But deciding under what circumstances to hold people, corporations, and others liable can be difficult.

Other elements are, of course necessary; no one wins a case just because he has sued the "legal cause" of his injury.[68] But causation is a necessary element of any civil tort lawsuit, and without that factor the plaintiff's case falls apart. Thus this chapter evaluates the central issues of causation.

The Conundrum of Causation

The problems with causation can be illustrated in a series of examples. Do we hold the bartender liable for the drunk driver's injuries (to himself or others) when everyone knew he drank 20 beers and then loudly an-

68. Other elements include a "duty" to the injured party and "damage" or injury caused by a violation of that duty. The element of "duty" can encompass a host of subsidiary issues. See generally [132]. Confusingly, the notion of "duty" is often used to test whether a cause of injury should or should not be the "legal cause" or legally responsible cause for that injury. This issue is properly discussed below in the context of foreseeable risk and multiple concurrent causation. But the "duty" element is not always coterminous with the scope of the foreseeable risk; in those circumstances, the existence of a duty is a separate element needed for liability. See [133]. Only the essential element of causation is discussed in this chapter.

nounced he was driving home?[69] When an aircraft crashes, we may hold liable the engine and airframe manufacturers, but how about the air traffic controller who saw the rainstorm on the screen and never issued a warning? When a car thief knocks someone down with his stolen car, do we hold the owner liable because he left the car unlocked and the keys in the ignition?[70] Do we hold tobacco companies liable for the cancer death of a smoker? Should we hold an armed bank robber liable for murder if an accomplice pulls the trigger in the heat of a robbery gone wrong? Should we hold a robber liable when a *police officer* shoots someone?[71]

More difficult are issues presented by a focus on alternative and concurrent causation. For example, two people fire the same type of weapon in the direction of the victim, and one bullet—we don't know whose—hits the victim and is fatal. Is either shooter liable? Both?[72] How about when a drowning results from the victim's inability to swim, someone's failure to properly supervise the children, another's failure to call for help, and someone else's kicking the victim into the water?[73] Ultimately, cases that appear to deal only with how far back to go in a chain of causation in actuality often present facts that suggest multiple concurrent (or nearly concurrent) causation.

From the difficult issue of concurrent causation, we turn to a more complex subset, where it is not clear when a force, person, or influence was the "cause" of injury. Who, if anyone, is liable when decades after the fact it turns out that a certain type of insulation kills people who used it?[74] Should *anyone* be liable when a farm pesticide, decades after it was ap-

---

69. Maybe. See *Williams v. Saga Enters., Inc.*, 274 Cal. Rptr. 3d 901 (Cal. Ct. App. 1990).

70. Sometimes, yes. See *Jackson v. Ryder Truck Rentals, Inc.*, 20 Cal. Rptr. 2d 913, 920 (Cal. Ct. App. 1993).

71. Yes, *People v. Caldwell*, 681 P.2d 274 (Cal. 1984). This recalls an earlier case in which the defendant shot his brother-in-law, *People v. Lewis*, 57 P. 470 (Cal. 1899) (the victim, realizing that he would die a long, slow, and very painful death, slit his own throat; the defendant was convicted of manslaughter for that death). See also *People v. Gardner*, 43 Cal. Rptr. 2d 603 (Cal. Ct. App. 1995) (compiling cases on "derivative liability for homicide").

72. Perhaps both. See *Summers v. Tice*, 199 P.2d 1 (Cal. 1948).

73. Any and all of these persons *may* be legally liable. See *Mitchell v. Gonzales*, 274 Cal. Rptr. 541 (1990), *aff'd*, 819 P.2d 872 (Cal. 1991). This is so even though any one of the causes alone would never have resulted in the drowning. See generally *Mitchell v. Gonzales*, 819 P.2d 872 (Cal. 1991); *Morgan v. Stubblefield*, 493 P.2d 465 (Cal. 1972); *Lareau v. S. Pac. Transp. Co.*, 118 Cal. Rptr. 837 (Cal. Ct. App. 1975) (car and train combined to kill passenger); *DaFonte v. Up-Right, Inc.*, 828 P.2d 140 (Cal. 1992) (combination of failure to warn, insufficient supervision, operator negligence, and poor design of equipment all led to serious injury to a 15-year-old's hand as he cleaned a moving conveyor belt of a mechanical grape harvester).

74. *Lineaweaver v. Plant Insulation Co.*, 37 Cal. Rptr. 2d 902 (1995) (asbestos).

plied to fields, is found to cause a statistical increase in the odds of getting cancer for those living in the area? Should the chemical company be liable? The farmers who applied it? The government agency that approved it? Suppose the only entity that knew the pesticide was dangerous happened to be a trucking company that shipped it—do they pay for the damage? Is the Coppertone suntan cream company responsible for millions of skin cancer cases because, for years, their advertisements encouraged people to tan their skin as dark as possible?

In this last set of difficult questions, the harm is a function of many factors. We may not be able to determine if the harm was caused by human or natural agencies; rather, the congruence of human, natural, and technical agencies caused the ultimate harm. Which one do we pick out as the *legally responsible* cause? Or should we blame all of the causal vectors and make every human and corporate actor responsible for the unanticipated injuries? We could, of course, blame no one, call the event an "act of God," and rest the causal "responsibility" with natural forces.

In the abstract, the law provides no good answers to these questions, because no *general* rule exists on how to pick, out of the infinite mass of all possible factors, the one (or two) on which legal blame is fixed. To illustrate, imagine that a factory burns to the ground when a match is lit and ignites the surroundings. Obviously the match, and the person who lit it, are the cause of the destruction. But assume this is a match-testing factory, in which for years billions of matches have been tested in a fireproof room filled with inert gases. A mistake is made, oxygen leaks in, and the factory explodes. We have the same causes at play, but we are tempted to identify as the "cause" not the match, but rather the gas leak.[75]

The Tort System's Causation Analysis

Cause in Fact

The traditional tort system first looks for "causation in fact" to resolve the conundrums posed by these examples. Some chain of circumstances that connects the accused's acts and the injury suffered may constitute a cause in fact, no matter how tenuous and no matter how many other contributing or intercepting forces there may have been.

That factual inquiry is essential, and it is worth noting that even theories of strict liability do not tamper with that essential inquiry. Strict

75. Compare *Merrill v. Los Angeles Gas & Elec. Co.,* 111 P. 534 (Cal. 1910) (escaping gas and light caused explosion) with *Lowenschuss v. S. Cal. Gas Co.,* 14 Cal. Rptr. 2d 59 (Cal. Ct. App. 1992) (gas company has no responsibility to purge gas from pipes and house meters in the path of oncoming fire).

liability, to be sure, dispenses with most classic elements of a tort, such as fault, negligence, or recklessness, or other measures of an accused's culpability [134]. But causation in fact is always required. This requirement, roughly approximated by establishing that the accused's acts were at least the "but for" cause of the injury, is a fundamental, intractable element of proof of a tort [135].

There are, to be sure, certain epoch-making cases that appear to undercut this requirement, but in fact they do not. For example, in *Summers*[76] only one of two shooters fired the wounding bullet; the court decided that unless one of the shooters could prove his innocence, both shooters would be liable. Plaintiffs ever since have tried to expand the rationale to other contexts in which they have difficulty pinpointing the actual cause.[77] True, a *Summers*-like doctrine may have the effect of holding liable one who actually was not the "cause in fact" of the injury. But it is important to note that the case simply switched the burden of proof on causation from the plaintiff, who is generally a person in a good position to know what happened to him, to the defendants who *in that case* were thought to be in a better position.[78] Causation in fact was still an element in *Summers*. The only difference from a typical tort case is that the defendants were saddled with the burden of proving its absence. As other cases have since noted, that switch of burdens of proof will not often be permitted.[79]

And then there are cases such as *Sindell v. Abbott Laboratories*.[80] *Sindell*, too, formally only shifted the burden of proof on causation. The court held that where some type of fungible good (the drug DES in this case) made by a plurality of potential defendants *in fact* caused injury to the plaintiff, then the odds that a given defendant's products caused the injury were equal to the percentage of the market for the fungible good held by that defendant.[81] That "market share" is then used to calculate the

76. 199 P.2d 1 (Cal. 1948).

77. See, e.g., *Lineaweaver v. Plant Insulation Co.*, 37 Cal. Rptr. 2d 902 (Cal. Ct. App. 1995). (*Summers* doctrine will not hold multitude of asbestos manufacturers responsible when an *undetermined* one was responsible.)

78. *Vigiolto v. Johns-Manville Corp.*, 643 F. Supp. 1454, 1457 (W.D. Pa. 1986). This switch of burdens is reminiscent of the venerable doctrine of *res ipsa loquitur*, which holds that when all of the many possible instrumentalities of injury are under the control of defendants, it is up to the defendants to prove their own innocence, rather than to the plaintiff to prove defendants' culpability. See, e.g., *Cooper v. Horn*, 448 S.E.2d 403, 405 (Va. 1994) (discussing *res ipsa loquitur*).

79. *Lineaweaver*, 37 Cal. Rptr. 2d at 906-08 (Cal. Ct. App. 1995).

80. 607 P.2d 924 (Cal. 1980).

81. *Id.* at 611–12.

percentage of the damages for which that defendant will be liable.[82] Crucially, *Sindell* acts only to shift the burden. Defendants still win when they can show that their products, as a matter of fact, could not have been responsible for the injury.[83]

Proximate Cause

But tracing back through a chain of causes is not enough for liability. Indeed, the confounding examples provided here *all* assumed that some chain of events in fact links the potential "cause" (such as a match lighting or the structural condition of an aircraft wing) with the harm (the exploding factory or crashed airplane). Quite aside from this causation "in fact," a wholly separate issue remains: legal or proximate cause. Proximate cause is the vehicle used by the law, in its wisdom and omniscience, to carry out society's policy by holding a certain agent liable. A court may hold some agents liable while excusing others who are also *in fact* linked to the disaster.[84] The doctrine of proximate cause is used to select, from all the causes in fact, the entity(ies) that will be held responsible for the injury.

This issue of *legal* responsibility, or *legal* causation, depends on the context. That is, do we look at just the match and its user? Or do we evaluate the larger context of the gases and atmosphere, and so perhaps sue the provider of the inert gases? Do we blame the materials of which the factory was built, with the possibility that the architects and builder are liable? Do we look at the entire idea of having a match-testing facility and suggest that it is inherently so dangerous that *anyone* who operates such a plant will be liable for any damages, without regard to negligence, fault or anything else? Do we sue the governmental entities that knew about the plant and did nothing to stop it?

The decision on which of the many contexts to use to evaluate liability depends on the sort of policy to be furthered by the liability rule. We presume that socially desirable conduct is furthered by singling out certain actors from the rest of the context, from the rest of all the other possible causes, and then punishing them as transgressors. As the canonical text on the subject states, a legal or "substantial" cause is that which "reasonable men" would agree is a cause, "using that word in the popular

82. *Mullen v. Armstrong World Indus.,* 246 Cal. Rptr. 32, 35 n.6 (Cal. Ct. App 1988) (asbestos not "fungible," so *Sindell* does not apply).

83. *Id.*; *Vigioltou v. Johns-Manville Corp.,* 543 F. Supp. 1454, 1460-61 (W.D. Pa. 1986); *In re Related Asbestos Cases,* 543 F. Supp. 1152, 1158 (N.D. Cal. 1982). See generally [136].

84. See generally *Maupin v. Widling,* 237 Cal. Rptr. 521 (Cal. Ct. App. 1987) and [137].

sense" [138].[85] Holding an entity liable is a statement, by the law and the culture it protects and in part defines, that those entities can and should bear responsibility for avoiding the harm at issue.[86]

Courts formally accommodate shifting public policy, changing mores, and technical developments by asking whether an injury was "reasonably foreseeable."[87] Reasonable foreseeability is essential for establishing proximate cause. One who reasonably should have foreseen a consequence, and was in a position to prevent it, may be liable for it. If an earthquake is reasonably foreseeable, then the architect may be liable for the house flattened by the tremor; if a flood is reasonably foreseeable, then the law imposes liability on the builders of the dam that fails.[88]

A fitting example of proximate cause's reliance on foreseeability is provided by the doctrine of *intervening* and *superseding* cause.

An intervening cause, as the term suggests, is simply one that intervenes between the defendant's act and the injury. When a contractor builds a weak foundation and an earthquake shakes it, loosening a brick and killing a nearby pedestrian, the earthquake is an intervening cause between the negligence and the injury. Courts decide if the intervening act was sufficiently *unforeseeable* so as to constitute a *superseding* cause—itself a wholly conclusory term—which would free the defendant of liability. Intervening criminal conduct may or may not be "superseding"; it depends on whether the court thinks the intervening criminal act ought to have been foreseen by the defendant.[89] Intervening "natural"

---

85. See *People v. M.S.,* 896 P.2d 1365, 1386-87 (Cal. 1995) (Kennard, J., concurring); *Mitchell v. Gonzales,* 819 P.2d 872, 882-85 (Cal. 1991) (Kennard, J., dissenting) (referring to the "social evaluative process" involved in deciding which of the infinite "causes" will be targeted for legal liability) and [137] ("Policy considerations underlie the doctrine of proximate cause").

86. See generally [139].

87. See, e.g., [140] (foreseeability as part of the analysis underpinning "legal" cause). While this chapter discusses only tort liability, it is worth noting that reasonable foreseeability also plays an important role in traditional contract law, in that breaching defendants normally are liable only for damages that were reasonably foreseeable, *Hadley v. Baxendale,* 9 Ex. 341 (1854); *Martin v. U-Haul Co. of Fresno,* 251 Cal. Rptr. 17, 23-24 (Cal. Ct. App. 1988).

88. *Cooper v. Horn,* 448 S.E.2d 403 (Va. 1994).

89. *O'Brien v. B.L.C. Ins. Co.,* 768 S.W.2d 64, 68 (Mo. 1989); see also *Doe v. Manheimer,* 563 A.2d 699 (Conn. 1989) (rapist's conduct deemed not reasonably foreseeable misconduct); *Erikson v. Curtis Inv. Co.,* 447 N.W.2d 165 (Minn. 1989) (a parking ramp operator owes customers some duty of care to protect against foreseeable criminal activity); *Akins v. D.C.,* 526 A.2d 933, 935 (D.C. 1987) (defendant computer manufacturer, whose negligence allowed a computer error that resulted in a criminal's early release from prison, held not liable for remotely foreseeable criminal assault).

forces such as floods,[90] power outages,[91] excessive rain,[92] wind,[93] and fog[94] may or may not be treated as acts of God for which no human agency is liable. Liability will not depend on whether humans were involved in some capacity (for they always are, in all of these cases), but only on whether the court believes that the action of the natural force should have been foreseeable.[95]

Because foreseeability is a matter of policy and perception, these things change over time. Bars once were not liable for drunks their bartenders watched drive away; now they may be. Landlords once were not liable for rapes on their premises; now they are in some cases.[96] Different

90. *Prashant Enter. v. State,* 614 N.Y.S.2d 653 (1994); *Saden v. Kirby,* 660 So. 2d 423 (La. 1995).

91. *Boyd v. Washington-St. Tammany Elec. Coop.,* 618 So. 2d 982 (Ala. Civ. App. 1993).

92. *Knapp v. Neb. Pub. Power Dist.,* No. A-93-134 1995 WL 595691 (Neb. App. Oct. 10, 1995).

93. *Bradford v. Universal Constr. Co.,* 644 So. 2d 864 (Ala. 1994).

94. *Mann v. Anderson,* 426 S.E.2d 583 (Ga. Ct. App. 1992).

95. The language in many of these cases mixes the test for causation *in fact* with that for *proximate cause.* Thus the courts' preliminary statements of the legal principles are often a muddle, although the actual evaluations and results in these cases are not necessarily wrong. For example, *Knapp* states that the defendant is liable "unless the sole proximate cause of that damage is an 'extraordinary force of nature,'" 1995 WL 595691 at *3. But this begs the analysis: the natural force (here, a rainstorm) will be held to be "solely" responsible (i.e., will be the "sole proximate cause") *if* the natural force was so extraordinary that the humans could not reasonably have foreseen it. Later, the court does in fact use that foreseeability analysis. *Id.* at *4 (citing *Cover v. Platte Valley Pub. Power & Irrigation Dist.,* 75 N.W.2d 661 (1956)). *Cooper* also states that the act of God doctrine applies only when the natural force "was the sole proximate cause of the injury...[and] all human agency is to be excluded from creating or entering into the cause of the mischief, in order that it may be deemed an Act of God," *Cooper v. Horn,* 448 S.E.2d. at 425 (citing *City of Portsmouth v. Culpepper,* 64 S.E.2d 799, 801 (Va. 1951)). Having found a human agency in the flood damage (a human-built dam that collapsed), *Cooper* decided that the act of God doctrine was inapplicable. *Id.* That analysis is flawed because there is always human agency as at least one of the contributing causes in fact. (The court was, however, probably influenced by its view that the human actions contributing to the damage were indeed foreseeable.) Human "agency" should be excluded as a consequence, not predicate, of proximate cause analysis. In contrast to *Cooper,* in *Mann* the act of God doctrine (which was also defined as excluding "all idea of human agency," 426 S.E.2d at 584) was held potentially applicable (thus relieving humans of all liability) to a series of automobile collisions in the fog—where the human agencies were absolutely obvious. But again, *Mann* correctly focused on the foreseeability of the ultimate injuries in the context of the fog and other weather conditions as presented to the defendants at the time of the accident. *Id.* at 585.

96. *Pamela B. v. Hayden,* 31 Cal. Rptr. 2d 147, 160 (Cal. Ct. App. 1994) (holding landlord liable for failing to provide adequate security in the underground garage of his apart-

views espoused by different judges on what should have been foreseeable have led not only to defendants' exculpation in favor of an act of God, but also to liability for natural catastrophes such as erosion, where defendants did nothing to cause the injury and there is not a whisper of evidence that they could have done anything to prevent it.[97]

What is "reasonably foreseeable," and so what qualifies as a "proximate cause," depends on custom and what people generally believe. These in turn may depend on general impressions of what technology can do. For example, it is "reasonably foreseeable" that shareware downloaded from the Internet may be infected with a virus—foreseeable to some, but not all. It is foreseeable that sensitive information transmitted via cellular phone and unsecured radio link can become public—foreseeable to some, perhaps, but not to a president of the United States and certain British royalty.[98] In each case, the courts do not focus on what individuals—even the individuals involved in the case—knew or expected, but rather on what should have been expected by a "reasonable" person, an abstract person compiled by a court from the whole population [140].[99]

What is deemed "reasonably foreseeable" depends on the court. Reading the tea leaves of contemporary society, different judges have different opinions. What is not reasonably foreseeable to one judge is perfectly predictable to another. A superseding event in one state may not be in another.[100] Reasonable foreseeability is a moving target; it dodges and weaves depending on public policy and on the perceived technological sophistication of the population. What is reasonably foreseeable depends on the sense of juries and judges, presumably reflecting what their culture believes is "reasonable." Thus the test acts as a mechanism by which the judicial system, in theory, remains responsive to its social environment. In the context of intelligent agents, the actions of the program or its progeny raise serious questions about the reasonable foreseeability, and thus causation, of the harms these programs will do.

---

ment building where a tenant was raped).

97. *Sprecher v. Adamson Co.,* 636 P.2d 1121 (Cal. 1981) (Bird, C. J.).

98. Former President Jimmy Carter reported to the White House "over an unsecured radio link" on his way back from a peace mission to Haiti and expressed surprise that his conversation had been recorded by others [141]. Some years ago, Prince Charles, too, apparently did not know that his cellular phone conversations with a mistress were public [141]. See generally *Rosh v. Cave Imaging Sys., Inc.,* 32 Cal. Rptr. 2d 136, 141-42 (Cal. Ct. App. 1994).

99. Citing [142].

100. *Pamela B. v. Hayden,* 31 Cal. Rptr. 2d 147, 160 (Cal. Ct. App. 1994).

CAUSATION IN THE DIGITAL DOMAIN

> *Once a computer system is designed and in place, it tends to be treated as an independent entity.*
>
> —Terry Winograd and Fernando Flores [143]

> *Diffusion and abandonment of responsibility*
> *[S]upervisors may eventually feel they are no longer responsible for what happens—the computers are.*
>
> —Thomas B. Sheridan [144]

Treating computers as responsible agents may mask the human authors of the mischief that may result from the use of computer systems [145]. Winograd and Flores provide the example of a medical diagnosis machine, which may be "blamed" for an erroneous diagnosis through plain *human* error (i.e., the system is used in the wrong context, or it is programmed with assumptions not shared by the users). The authors suggest that those human users and programmers, not the machine, are responsible [145]. However, in the complex processing environment envisioned in this chapter, one cannot impose liability on any identifiable agency, human or otherwise.

This is the case because multiple agent systems imply at least the causal input of multiple independent programmers of the basic scripting or authoring software—a vast number of users creating distinct intelligent agents and an unpredictable number of agent-to-agent interactions on an unpredictable number of interwoven platforms, operating systems, distributed data programs, and communication programs, each of which in turn incorporates at least some further limited programming. This inevitable causal complexity poses problems for traditional tort law, in which a determination of proximate cause is essential, as it evaluates the liability of an intelligent machine system.

An Example of an Intelligent Processing Environment

As we illustrate an intelligent system, we recall the comments of researchers quoted earlier: "We envision a world filled with millions of knowledge agents, advisers and assistants. The integration between human knowledge agents and machine agents will be seemless [sic], often making it difficult to know which is which" [146]. This chapter assumes that intelligent agents will be created both by humans and by other intelligent agents, as needed.[101] Humans can now use programs such as Telescript and Java[102] to make intelligent applets.[103] These agents, created at diverse computers, will communicate with each other to accomplish larger goals. Even now small programs (fondly referred to as daemons) stand ready to assist those involved in telecommunication sessions. They can inform the calling system that a connection is available, receive a request from a remote user, and service the request by shifting directories, sending data and so on [149]. Other programs such as Java know how to manipulate these daemons, and we can expect that all of these programs will know how to interact with other programs such as Internet search engines,[104] graphical web browsers such as Netscape, intelligent natural language query tools, and a host of other useful programs resident in one machine or another. No one of these small programs could accomplish the ultimate goals of a human user, but together their actions may *in the aggregate* appear intelligent in a broad sense.

The Structure of Alef

To illustrate some attributes of the processing system in which intelligent agents will work, imagine a hypothetical intelligent programming environment that handles air traffic control, called "Alef."[105] The description

101. In the electronic arena, humans and software communicate in the same way, and when both are capable of spawning software agents, it is likely that others in the electronic medium will be unable to distinguish between human- and machine-originated communications. See generally [147]. Consistent with the general model of intelligent agents mapped out in this chapter, Maes describes agents [148]: "[Agents are] distributed, decentralized systems consisting of small competence modules. Each competence module is an "expert" at achieving a particular small, task-oriented competence. There is no central reasoner, nor any central internal model. The modules interface with one other via extremely simple messages.... As a result the behavior produced is robust, adaptive to changes, fast and reactive."

102. See generally [91].

103. Recently, software agents were "evolved" by other computer programs in an apparently successful effort to create intelligent autonomous agents that would assist in securing computer systems from unauthorized attacks [149].

104. ARCHIE, VERONICA, and GOPHER are three euphemistically named search engines that reside at various systems throughout the world, ready to accept requests from a series of other remote programs (or, lest we forget, from humans as well) to search the Internet for selected items.

105. See [151]. Air traffic control presents a series of inherently concurrent computational problems, and thus it is well suited to the use of a large collection of concurrently active agents [152]. A wide variety of uses exists for intelligent agents in other distributed systems, such as for information management [153,154], manufacturing [155], air combat and related situational awareness contexts [156], and telecommunication [157].

that follows emphasizes the networked distribution of agents, their unpredictable variety and complexity, and the polymorphic ambiance of the intelligent environment as a whole.[106]

Alef's fundamental task is to guide the flow of air traffic in and out of a section of space, including airports in the area. Today, this work is in fact accomplished by thousands of intelligent agents (both human and computer) in airplane cockpits, airport control towers, air traffic control centers, and weather forecasters. There are a number of modules, or agents, that would be a part of any large-scale intelligent processing environment. Alef is an example of a large-scale processing environment composed of many modules. Those modules within Alef are given here in broad, sometimes overlapping strokes, with examples specific to Alef's tasks. Many of these modules would in fact be composed of discrete submodules.[107] Examples of specific tasks within Alef which would be performed by such submodules include:

- *Sensory input:* Alef will use voice, visual, and motion information, as well as traditional data input. Voice- and motion-recognition software, as well as software to accurately read and analyze digitized visual input, are complex and sophisticated. Each contains subsidiary modules. Alef would secure information from airport surface sensors, radar, and human voices.
- *Sensory control:* Alef may need to turn on and off lights, open and close microphones, modify the sensitivity of and/or select various sensors, and undertake robotic actions.[108]
- *Data input:* Alef will need to utilize text files, knowledge bases, [161], and other databases. Alef should know the "rules of the road" such as the federal aviation regulations, facts about the aircraft it is guiding (such as speed, position, fuel, load, range, altitude, destination, and origination, as well as aircraft capabilities such as maximum rate of climb and engine-out glide distances), and weather information. The data reside in hundreds of physically scattered sites.
- *Heuristic analysis:* Alef should have access to expert systems, rules of thumb, and heuristics. These would, for example, recommend

106. Of course, the ensemble of intelligent agents contemplated by this chapter is not operational, and so the outline here is speculative.

107. Professor Sheridan has a useful list describing many of the precursors to the automated systems listed next. He describes modern autopilots and associated programming functions, on-board collision-avoidance systems, and the like [158]. Many sources, such as advanced pilot's manuals, provide information on the human and machine systems currently controlling the nation's airspace. See, e.g., [159,160].

108. That is, the physical movement of machines linked to Alef.

separation between various types of aircraft (over and above the minimum requirements of FAA regulations) and which approaches, departures, and runways to use depending on weather, general traffic, and facts specific to a given aircraft (such as its destination).

- *Basic assumptions:* Alef should have assumptions about the way the world works—a grasp of gravity and that two airplanes cannot inhabit the same space are the most obvious.
- *Goals and commitments:* Alef may be equipped with certain goals, such as minimizing the rate of fuel burn and the amount of flight time en route. Alef should possess a preference for rapid responses to certain types of urgent communications from aircraft. Also, Alef may be committed to neighborhood noise abatement involving difficult or circuitous approaches. Or it may prefer easy or simple instrument approaches to those involving many way-points, complicated turns, or holding patterns.
- *Basic computing, load-sharing and load distribution:* Alef must have the capability to distribute its processing requirements and then integrate the results. This is not a trivial task. Alef must constantly manage data that are processed by a variety of distributed programs, in the context of rapidly changing demands for processor and memory resources in different locations.
- *Resource management:* Alef must have power management to ensure continuous operations in emergency situations and guaranteed access to the communications networks, such as telephone, cable, satellite, and microwave channels, to ensure required communications among agents. Because it is essential to keep Alef operational, Alef may have agents devoted to preemptive or dedicated access to power and communications facilities.
- *Programming:* Alef may find it expedient to make new subroutines. This would have the effect of modifying the substance, or the actions, of other agents. Obviously, agents useful only under certain circumstances (e.g., in low-visibility conditions) could be turned off altogether when those conditions did not obtain; other agents might be modified, such as those controlling power requirements, those that deal with information overload, or those that adjust the intensity of runway approach lighting in reaction to weather conditions. At a lower computational level, Alef would also manage its memory and other systems tasks.
- *Communications:*[109] Alef should be able to accomplish routing of messages (data), registration (address identification) of the other agents in the environment, and mediation and translation of mes-

109. See [162] for a discussion of protocols for exchanging information and knowledge.

sages among the various agents. Alef would also communicate with other intelligent systems controlling aircraft within their jurisdictions across the United States and abroad.

- *Interface:* The intelligent environment will interact with humans. Some of these agents may also be categorized as sensory input agents. Alef will use graphical user interfaces and other programs to communicate with humans on the ground and in the air [163]. Thus, Alef will often communicate with human pilots, although in emergency situations Alef may directly control the flight of an aircraft itself.

The Operation of Alef

Routine Operations. Alef encompasses the plurality of agent types outlined here. Each agent is connected to a variety of similar agents, as well as to many wholly distinct agents. Agents run in—and transmit themselves between—a variety of locations. In the Alef example, agents may be found on board hundreds or thousands of aircraft, on vehicles and other sites on airport surfaces, in satellites, at control towers, and in more centralized computing systems. Agents are continuously added into and taken out of the mix. The sources or architects of these agents vary widely. Some agents will have been written by aircraft and radar manufacturers, some by pilots programming their own aircraft,[110] some by certain of Alef's own agents, and others by human operators in control towers, at air traffic control centers, and at weather observation facilities. The data streams into Alef will be truly staggering, including changeable weather and highly variable data on each aircraft in Alef's space or about to enter its space. Managing this complexity of data is precisely the rationale for Alef's existence.

Alef routes air traffic. In emergencies, as Alef perceives them, Alef may control individual aircraft to prevent collisions. That, and Alef's more mundane actions, would be a function of the totality of the agents (both currently active and previously active) which modified data or which modified other linked agents.

110. An elementary example of this used today is the "programming" by pilots of their transponders, devices that respond to usually ground-based radar queries with a code selected by the pilot. Some codes indicate emergencies; some mean the aircraft's radio is out of commission; others indicate an aircraft flying under visual (as opposed to instrument) flight rules. Pilots also individually "program" or set their altimeters, information from which is passed via radio to ground-based controllers and at times to other aircraft to determine if a collision is imminent. Pilot-created programs also navigate via selected way points at given altitudes, and so on.

To what appears to a human to be repetitions of the same problem, Alef may arrive at different responses. For example, when guiding an arriving jet, Alef may route it over one way point at one altitude one day and by a different route the next day. Communication processing and other tasks would differ from day to day and from moment to moment. Mandating one approach or the other would defeat Alef's purpose.

Pathological Operations. Alef's pathological operation cannot specifically be forecast. But if Hofstadter is right [164], some pathology will erupt as a function of Alef's attempt to solve a new problem by analogy to an old one. The analogizing process will require Alef to combine its constituent distributed agents in new and interesting ways.[111] The result will be an unexpected configuration or interaction of agents, including perhaps new agents created to handle the new problem. An anomalous situation may cause an anomalous result; based on its prior experience, Alef may decide to ignore certain data and attend to new information. Alef may use the "wrong" analogy to incorporate data. For example, a rapid decrease in an aircraft's altitude may be seen as evidence of fuel problems and not the near-miss air collision it was. In short, Alef may be seen in retrospect as having exercised poor judgment.

Specifically, Alef may ignore certain warnings. For example, noting that ground or collision proximity warnings activate long before a collision is actually imminent, Alef may route traffic closer to obstructions than it should. Perhaps in its enthusiasm to ensure continuous power, Alef reroutes electrical power away from the surrounding homes. In an effort to streamline operations, Alef may encode data under its control, effectively making it unreadable to humans and useless for other purposes. Perhaps the system misunderstands communications from a pilot or from instruments on board an aircraft, or assumes the pilots have information they do not, and as a result a plane runs out of fuel. The specific admixture of agents that generates the poor judgment is more likely to occur at a lower, and less overt, computational level. Conflicting claims upon its computing or memory management resources may cause Alef temporarily to drop communications with an aircraft at a moment that turns out to have been critical.

Of course, each of these problems can be fixed. Alef can simply be overridden with instructions never to deviate from preassigned clearances between aircraft, never to modify its access to electrical power, never to tamper with the format in which its data is held, and so on. Global constraints can easily be imposed, once we know of the problem.

111. See generally [1].

These will prevent systems from "evolv[ing] in ways independent of the creator's intent" [111].

But we do not know the problems in advance. Experience is the great teacher, and sometimes the experience will come first to the intelligent system, *before* humans can accommodate the crisis. Global constraints defeat the purpose of intelligent systems such as Alef. To be sure, these constraints are a matter of degree.[112] However, the more global constraints control the decision-making ability of the system, the less the system's intelligence is a function of its ensemble of collaborative agents. The environments discussed in this chapter do *not* have a "central reasoner."[113] For a Hofstadterian program, the guiding analogy is not presented in advance; the intelligent system detects or invents it. Nor do these systems exist aside from a larger networked environment. Thus, it is meaningless to suggest, perhaps as another form of global control, that the system be severed from a network and neatly encapsulated into a stand-alone box. The imposition of such global constraints would not simply modify the behavior of these intelligent systems, but would eviscerate them.

Unpredictable Pathology: Absolving Humans of Liability

If and when Alef, as an entire processing system, "causes" an unanticipated fault, it will be difficult to establish human liability. For it is not enough that some undifferentiated damage of *some* kind may be expected. We "expect" those problems, surely, from many software environments.[114] Rather, liability attaches to those who reasonably should have foreseen the *type of harm* that in fact results. That is how the "reasonable foreseeability" test is implemented. Liability does not extend to the "remarkable" or "preposterous" or "highly unlikely" consequence [166–168].[115] Of course, this will always be a matter of degree. Consequences may not be *so* unforeseeable as to relieve the causative agents of responsibility [170]. As noted earlier, the bounds of foreseeability fluctuate, varying from court to court and over time.

---

112. See [165] (discussing supervisory control of automation).

113. See [147].

114. See above, the discussion of the unreliability of software; and see also Chapter 7, "Information Loss and Implicit Error in Complex Modeling Machines."

115. Reference [167] notes that results should not be extraordinary for liability to attach. See also [169], which points out that there is no liability for acts of God if those acts are extraordinary and bring about an injury different in kind from that originally threatened by defendant's act.

The inability to pinpoint specific human responsibility for failure suggests that "the machine" or the network "system" should be blamed for damage it causes. The temptation to treat sophisticated intelligent agents as independent legal entities, thus absolving the humans involved, is powerful. The agents appear to be autonomous and independent; their pathological results are by definition unpredictable. No human will have done anything that *specifically* caused harm, and thus no one should be liable for it. Just as we are not liable for the consequences of a human agent's unforeseeable pathological actions,[116] so too humans should be absolved of liability for the unforeseen results of machine intelligence's pathology.

Humans are, of course, held legally responsible for some computer malfunctions. As noted in the discussions of classic expert systems, many "computer errors" are easily traced to human negligence or other fault. It makes good sense to impose liability on those persons who have the power to avoid the injury in the future.[117] But that rationale for imposing liability fails when no particular human has the ability to prevent the injury, short of banning the use of intelligent agents altogether.

Liability in the computer context must depend, as it does in other contexts, on plaintiffs' ability to convincingly argue that a given injury was reasonably foreseeable. This means foreseeing the context of an action (or inaction), and being able to predict at least in general terms the consequences of the action on the context.[118] But even today it is often difficult to charge people with predictive knowledge of the electronic context and of the consequences of various deeds on that context.[119] For example, a hard drive's (programmed) controller works well in most circumstances, and generally no data is lost. It turns out that with the advent

116. See *Lopez v. McDonald's*, 238 Cal. Rptr. 436, 445-46 (Cal. Ct. App. 1987) (holding McDonald's not liable for deaths of plaintiffs caused by an unforeseeable mass murder assault at its restaurant).

117. Compare *State v. White*, 660 So. 2d 664 (Fla. 1995) (police department's failure to update its computer records for days after an arrest warrant was served caused second unjustifiable arrest; court suppressed evidence seized at second arrest) with *Arizona v. Evans*, 115 S. Ct. 1185 (1995) (no suppression of evidence when computer was negligently maintained by court officer *outside* of police department). The suppression rule is designed to encourage the police to act properly: the Florida Supreme Court argued that policy is furthered by suppression in the Florida case and not in the Arizona matter. See also *Bank Leumi Trust Co. v. Bank of Mid-Jersey*, 499 F. Supp. 1023 (D.N.J. 1980) (bank liable for computer's failure to read handwritten notation on a check; not an act of God).

118. Courts do not require that the person held liable actually have undertaken this forecast. Rather, the courts hold that if it was reasonable to have expected that forecast, the defendant will be treated as if the forecast had been made.

119. See [171].

of 32-bit operating systems such as IBM's OS/2 Warp, an attempt to process multiple interrupts causes the controller to occasionally corrupt data stored on the drive with possibly catastrophic consequences. Is this a case of a "flawed" controller, as the media suggests? Or is this a flawed operating system? After all, not all 32-bit operating systems cause this data corruption [172].[120] Perhaps both the maker of the drive controller and of the operating system are eligible for "blame" here. Yet no "reasonable" person would have predicted the cause of the failure, so there is no basis on which to hold a given manufacturer liable. The problem is simply unforeseeable incompatibility in the linked systems.

This problem becomes more acute with artificial intelligences such as Alef, which operate in environments in which elements are continuously added and dropped out. The specific pathological judgment calls made by Alef are not reasonably foreseeable, and thus courts should treat them as superseding causes, corresponding to unexpected fog or storms which, in former "natural" contexts, eluded human responsibility.

Failure of Causation: Absolving Programs of Liability

We are beginning to work in a class of fundamentally non-linear media...[and] the concept of logic pretty much goes out the window. What's going to replace it?

—John A. Rheinfrank, quoting John Brown,
Chief Scientist at Xerox's PARC [173]

Fixing liability on networked systems will solve no problems. Social policies are not furthered when courts decide to blame an ensemble of networked agents. The law does not yet recognize programs as legal entities capable of defending themselves or paying damages.[121] Targeting intelli-

---

120. The USENET group comp.risks has reported on this example of unpredictable conflicts erupting between operating systems and symbiotic hardware components: "Intel and other computer companies are trying to determine the extent of problems caused by a flaw in an RZ-1000 EIDE controller chip that is included in some early PCI motherboards manufactured by PC Tech. The flaw was discovered in 1994 and was corrected through a software 'patch,' but the latest version of [IBM's operating system] OS/2 disables that patch." See RISK-LIST: Risks-Forum Digest, Aug. 18, 1995, available from the Internet USENET group comp.risks at http://catless.ncl.ac.uk/Risks (citing *Investor's Bus. Daily*, August 16, 1995, at A15).

121. That is the case now, but I have outlined a framework for treating electronic personalities as fully fledged legal entities [174]. That framework does not, however, address or solve the problems discussed here of distributed agency and the breakdown of causation.

gent agents for legal liability fails to solve two fundamental and interrelated problems. First, interacting data and programs "responsible" for a failure are as distributed as the involved human agencies. Second, classic cause-and-effect analysis breaks down.

The dispersion of the agencies involved—human and otherwise—has been described earlier.[122] These agencies do not simply trigger each other sequentially, in a specific time or space order. We may have multiple concurrent causes, but there is no mechanism for selecting out those which are legally "substantial" (and hence lead to liability) from those that are in some fashion incidental. Over time, the configuration of active elements shifts: different agents act on mutating data and different sets of agents interact from moment to moment. In an eternally changing context, agents have no inherent substantiality or persistence. They are polymorphic. The agents' roles change from centrally active, to sustaining context, to inactive or absent altogether from the processing environment.

The notion of "proximate" or "legal" causation implies a court's ability to select out on a case-by-case basis the "responsible" causes. But where damage is done by an ensemble of concurrently active polymorphic intelligent agents, there is insufficient persistence of individual identifiable agencies to allow this form of discrimination. In this context, there will generally be no room for the courts to use social policy to distinguish among the programs and operating systems, encapsulated data, objects, and other data-cum-program entities—each of which is a cause in fact (and perhaps a cause *sine qua non*) of the injury or damage.

On what should society and the courts focus? A court may have a felt sense that it should hold the manufacturers of exploding vehicles responsible for resultant injuries, even when there are a series of other eligible causes (cars are driven too fast, tank manufacturer made the gas tanks out of thin metal, etc.). Likewise, a court may hold accountants and lawyers responsible for allowing a company to issue a false securities prospectus because they should have foreseen the injury. A court makes choices in an effort to fix blame where it feels a need to compensate the injured and to encourage desirable behavior by those able to foresee the harm they may cause. But where "fault" is a function of many ephemeral agents' unpredictable interactions, there is no target for the judgment of social policy.

What I have termed pathological outcomes are *inherent* in the electronic cosmos we are making. Pathology is both "immanent and elusive from an excess of fluidity and luminosity" [176]. "[V]ery strange answers" [1] are exactly the sort of responses we expect from our agents, and it is unlikely that we will ever arrive at a stable global electronic ecology that

122. See generally [175].

buffers us from these strange answers. Indeed, it is likely that networks will become increasingly linked, reactive, and complex, and so therefore increasingly entrusted to the day-to-day management by agents. In this sense, error is emergent, caused by all and none of the transitory participating elements.

No surgery can separate these inextricably entwined causes. No judge can isolate the "legal" causes of injury from the pervasive electronic hum in which they operate, nor separate causes from the digital universe that gives them their mutable shape and shifting sense. The result is a snarled tangle of cause and effect as impossible to sequester as the winds of the air or the currents of the ocean. The law may realize that networks of intelligent agents are not mysterious black boxes, but rather are purposeful, artificial constructs. But that will not solve the problem of legal liability. The central doctrine of proximate cause, essential in the sorting out of multiple causes and tagging some in accordance with public policy, is useless when causes cannot be sorted out in the first place.

TURING, WHERE ANGELS FEAR TO TREAD

The electronic universe is multiplying at an exponential rate, absorbing many segments of the infrastructure, and at least indirectly controlling much of the rest. In that utterly dispersed place, traditional doctrines of legal liability, such as cause and effect and foreseeable injury, do not exist. This will shock not only lawyers and judges, but also the businesses that expect to be compensated for their efforts and the injuries they suffer. If we try to use a system based on sorting through various causes to navigate a miasma where "cause" means so little, the legal system and the rapidly developing technology will suffer. People will be unfairly charged for damages they could not have prevented or previewed. Extravagant laws passed by legislators unclear on the technology will scare developers, systems operators, and resource providers from fully engaging in the commerce of the electronic universe. If there were an alternative that plainly recognized the impotence of the traditional tort system, we might avoid some of these frantic and destructive efforts to control the uncontrollable. To that end, I propose the Turing Registry.[123]

123. Apostles of the master will recognize the source [177]. Alan M. Turing was a brilliant British mathematician who devised in 1936 the formal architecture of the computer, an abstraction termed a "Turing Machine" [178]. The "Turing test" is one putatively for intelligence, perhaps conscious intelligence, in which conversational responses of hidden humans and machines are passed via console to human testers. If the testers cannot distinguish the human from the machine, the machine is said to pass the Turing test. There is little agreement on whether the test has ever been passed, or indeed is meaningful. See, e.g., [179,180].

The Registry

The behavior of intelligent agents is stochastic. Like risks underwritten by insurance agencies, the risk associated with the use of an intelligent agent can be predicted. This suggests insuring the risk posed by the use of intelligent agents. Just as insurance companies examine and certify candidates for life insurance, automobile insurance, and the like, so too developers seeking coverage for an agent could submit it to a certification procedure, and if successful would be quoted a rate depending on the probable risks posed by the agent. That risk would be assessed along a spectrum of automation: the higher the intelligence, the higher the risk, and thus the higher the premium. If third parties declined to deal with uncertified programs, the system would become self-fulfilling and self-policing. Sites should be sufficiently concerned to wish to deal only with certified agents. Programmers (or others with an interest in using, licensing, or selling the agent) would in effect be required to secure a Turing certification, pay the premium, and thereby secure protection for sites at which their agents are employed.

An example may help here. Assume a developer creates an agent that learns an owner's travel habits and can interact with networked agents run by airlines, hotels, and so on. The agent is submitted to the Turing Registry, which evaluates the agent's risk. How likely is it to interact with agents outside the travel area? Does it contain a virus? How responsive is it to remote instructions? How quickly does it cancel itself when launched into another network? Does the agent require supervisory control, and to what extent? What decisions can it make on its own? Generally how reactive is it? Agents designed to interact or capable of interacting with nationwide power grids or nuclear power plant controllers would be assessed a correspondingly high premium. In all cases, the premium would be a function of the apparent reactivity of the program, its creativity, and its ability to handle multiple contexts—in short, its intelligence. High intelligence correlates with high risk. Such intelligence and corresponding risk is associated with agents designed to interact with a wide variety of other agents and which, concomitantly, are not designed to submit to global controls [58].[124] The Turing Registry predicts the risk posed by the agent and offers certification on payment of a premium by the developer.

124. See above, the discussion of Alef's pathological operations.

The Registry will certify the agent by inserting a unique encrypted[125] warranty in the agent.[126] United Airlines, Hilton Hotels, travel agents, and so on would express trust in the Registry by allowing into their systems only Registry-certified agents. The encrypted warranty may also be used to ensure that certified agents only interact with other certified agents.

At some point a pathological event will occur. Perhaps United's database is scrambled. Perhaps an intelligent agent, trying to do its best, finds out how to reroute aircraft to suit a trip request, or it blocks out five floors of a hotel during the Christmas season—just to play it safe.[127] At that point the Turing Registry pays compensation if its agents were involved, without regard to fault or any specific causal pattern. That is, if the Turing Registry determines that a certified agent was involved in the disaster as a matter of cause *in fact*, the Registry would pay out. No examination of *proximate cause* would be undertaken. The Registry would provide compensation if and when Alef tampered with data in linked systems. And without an impossible investigation into the specific agents directly responsible, the Registry would pay out if an aircraft under Alef's control was diverted into a mountain. The Registry might pay for mid-air collisions, the consequences of data and program modules residing in computers owned by the government at air traffic control centers, by private pilots and by companies providing various navigation and data-management services.

Risk might be lowered with minimum requirements. Because agents' environments are just as important as the agent itself,[128] perhaps the Registry may not pay compensation unless at the damaged site (1) certain environmental controls are in place, designed to curb agents and limit their reactivity, and (2) essential security, data backup, and other redundancy systems are in place. Perhaps the Registry would insist that covered sites *only* allow in registered agents. But, as discussed earlier, the imposition of substantial global controls will defeat the purpose of the intelligent system, and thus such requirements offer very limited protection.[129]

---

125. Programs such as Phil Zimmermann's Pretty Good Privacy® (PGP) can do this now. Encryption software can provide the means to authenticate a digital message as from a given source. PGP is available as freeware. See [181–189].

126. Agents may mutate, and accordingly the certification must be able in some sense to "follow" the agent through its incarnations. This issue is discussed below under the section, "Replicant Sampling and the Revenge of the Polymorph."

127. Hotels do in fact link their reservation systems to the Internet. See note 19.

128. See generally [80]. In what I have dubbed a *processing environment* it is not possible to separate environment from programs.

129. Limiting the access, reactivity and (in effect) the scope of an agent's judgment doubtless will limit the harm agents can do. However, this will come at the cost of eviscer-

Insurance companies already exist, as do companies that test programs for compatibility with a variety of hardware and software. Some recently formed companies specializing in electronic security will probably escrow the electronic *keys* needed to access encrypted data [190]. Others can authenticate data, or digital signatures, which unambiguously identify an author or source [191]. New companies are being formed to allow secure financial transactions on the Internet.[130] Expertise from a combination of such companies could be harnessed to provide the type of secure, imbedded certification described here.

A Comparison to Traditional Insurance

Most insurance schemes are based on traditional tort causation analysis.[131] A brief review of the various types of traditional insurance reveals their ultimate dependence on classic proximate cause analysis. These schemes' reliance on proximate causation is the fundamental reason why the Registry cannot function like—and therefore is not—an insurance scheme in the traditional sense.

Insurance usually just redirects, or spreads the risk of, an award of damages for a tort. Similarly, doctrines of vicarious liability such as master–servant and parent–child implicate identical issues of tort liability, the only difference being that the person paying is not the negligent actor.

Thus third-party (or liability) insurance pays out when someone else establishes a tort against the insured;[132] workers' compensation pays out when injury occurs in the workplace, when otherwise the employer might have been held to be responsible.[133] First-party insurance covers losses sustained directly by the insured.[134]

---

ating agents' ability function as distributed intelligences (e.g., their ability to manage the complex electronic province entrusted to them).

130. CyberCash, Inc., enables encrypted transmissions of credit card information across the Internet [192].

131. See [193] ("The analogy between insurance and tort cases on issues of proximate cause is quite close").

132. See, e.g., *Fireman's Fund Inc. Co. v. City of Turlock,* 216 Cal. Rptr. 796 (Cal. Ct. App. 1985); *Int'l. Surplus Kines Ins. Co. v. Devonshire Coverage Corp.,* 155 Cal. Rptr. 870 (Cal. Ct. App. 1979).

133. Cal. Ins. Code § 109 (West 1993) ("[I]nsurance against loss from liability imposed by law upon employers to compensate employees and their dependents for injury sustained by the employee arising out of and in the course of employment, irrespective of negligence or of the fault of either party").

134. *Garvey v. State Farm Fire & Casualty Co.,* 770 P.2d 704, 705 n.2 (Cal. 1989) (contrasting first- and third-party policies).

Some insurance polices are "all risk," and broadly insure against all losses described in the policy unless specifically excluded.[135] Broadly speaking, the Turing Registry would offer a form of insurance, as it would "indemnif[y] another against loss."[136] But traditional causation, including its proximate cause analysis, is a fundamental underpinning of insurance coverage schemes.

Thus, for example, most policies will *not* cover losses caused by the "willful act of the insured."[137] While there is debate on the meaning of this willfulness exclusion, it has barred coverage for losses caused by the insured's intentional crimes,[138] for wrongful termination,[139] and for various "intentional" business torts.[140] As long as insurance companies can invoke exclusions such as these, the classic issues posed by proximate cause analysis will exist. More generally, even all risk, first-party insurance will exclude perils or excluded risks, which will generate conflict between insureds and insurance companies on whether the excluded risk was or was not the "proximate cause."[141] Thus both traditional third-party liability policies and first-party policies assume an intact, fully functioning proximate cause doctrine. Therefore, neither of these insurance systems can address the liability of distributed AI.

The Registry's Limitations

There are at least two practical problems with the sketch provided here of the Registry's operations, and perhaps a third theoretical problem. First,

---

135. *Strubble v. United Servs. Auto. Ass'n,* 110 Cal. Rptr. 828, 830-32 (Cal. Ct. App. 1973).

136. See Cal. Ins. Code §§ 22, 250 (West 1993) (defining "insurance").

137. Cal. Ins. Code § 533 (West 1993) (forbidding coverage for willful acts by the insured). See also Cal. Civ. Code § 1668; *Clemmer v. Hartford Ins. Co.,* 587 P.2d 1098, 1105-06 (Cal. 1978).

138. See, e.g., *State Farm Fire & Casualty Co. v. Dominguez,* 182 Cal. Rptr. 109 (Cal. Ct. App. 1982) (first-degree murder).

139. See, e.g., *St. Paul Fire & Marine Ins. Co. v. Superior Court,* 208 Cal. Rptr. 5 (Cal. Ct. App. 1984).

140. See, e.g., *Aetna Casualty & Sur. Co. v. Centennial Inc. Co.,* 838 F.2d 346 (9th Cir. 1988); *State Farm Fire & Casualty Co. v. Geary,* 699 F. Supp. 756 (N.D. Cal. 1989); *Allstate Ins. Co. v. Interbank Fin. Servs.,* 264 Cal. Rptr. 25 (Cal. Ct. App. 1989).

141. *Garvey v. State Farm Fire & Casualty Co.,* 770 P.2d 704 (Cal. 1989). See also *LaBato v. State Farm Fire & Casualty Co.,* 263 Cal. Rptr. 382 (Cal. Ct. App. 1989). *Garvey* used the phrase "efficient proximate cause" in the sense of the "predominating cause" as distinguished from some "initial" or "moving" causes, which would give an undue emphasis on a temporary prime or "triggering" causation, *Garvey,* 770 P.2d at 707. Thus *Garvey* applies proximate cause in roughly the traditional way. See generally Cal. Ins. Code §§ 530, 532 (West 1993) (causation requirements for coverage).

the Registry will not preempt lawsuits directed at the scope of coverage provided by the Registry or lawsuits by individuals outside the system. The second problem is created by the technical difficulties in checking the reliability of agents. The third problem concerns the perceived difficulty of identifying the source of an agent or its code that causes damage in fact. A discussion of these will illuminate the proposed operation of a certification registry and some of its limits.

Coverage Disputes and Victims Outside the System

First, there is some question whether the model sketched out here will really remove from the legal system disputes attendant on damage done by AIs. There is presently much litigation involving the scope of ordinary insurance coverage, and the model presented here would not necessarily preempt such lawsuits among (1) the Registry, (2) programmers of agents, and (3) parties operating site hosts (such as United Airlines and hotels in the earlier example) who allow in certified agents.

Generally, though, these will be relatively ordinary contract disputes, not cases requiring an impossible cause-and-effect analysis of the damage done by intelligent agent ensembles. Thus, the usual sort of contract (policy) lawsuit might be brought by damaged sites if the Registry failed to pay. Programmers might allege illegal restraints of trade if the Registry declined to certify certain agents. A site owner damaged by an agent designed to cause the damage could still sue the programmer of that agent for the willful injury.

Furthermore, the Turing system sketched out here may not have the ability to handle damage done when, for example, Alef preempts a power supply and browns out the neighborhood or when aircraft fly too low, creating a nuisance in the area. Distributed AI systems are likely to have consequences *outside* our computer networks, in physical contexts that cannot be "certified" by a Registry. But this issue, too, should be of decreasing importance over time. If the assumptions of this chapter are correct, complex digital systems will increasingly ramify throughout the infrastructure.[142] Thus, Alef's assumed control of the power grid might well be a function of the AI's authorized reception by a machine host. If so, damages caused "in fact" by that infiltration would be compensable.

Unresolved by this discussion is the extent of the Registry's reimbursement coverage. There is a spectrum of potential damage ranging from direct injury, such as data corruption, to the immediate effects of a misjudgment, such as mid-air collision, to less direct effects, such as

142. See the introduction of this chapter.

neighborhood brownouts, to highly indirect and consequential effects, such as a decline in the stock market value of a company that makes aircraft or radar parts. Presumably the Registry could, by contract, specify that compensation would be paid only for immediate pathological effects (i.e., for damage caused to networks and mechanisms immediately controlled by networks). Registries might also simply limit their payout on a per-agent, per-site, or per-event basis.

Technical Difficulties in Checking Software

There are problems with the technical methods to be used to check or analyze intelligent agents. We should assume the problems in verifying the actions of tomorrow's agents, and of ensembles of agents, to be more difficult than those involved in today's complex software environments.[143] However, no registry—Turing or its competitors—needs to establish the eternal reliability of agents. To the contrary, this chapter assumes that the agent ensembles are not always reliable. The issue is rather whether an examination of agents may reveal the *likelihood* of unpredictable and dangerous behavior over the long run. That is a technical question beyond the scope of this chapter, but success with that sort of stochastic evaluation—perhaps with the criteria I suggest earlier, such as reactivity and environmental safeguards—is not inconsistent with the theoretical difficulties in predicting the specific behavior of complex programs.

Replicant Sampling and the Revenge of the Polymorph

:10000000800900077470652E41534D188820000048
:10001000001C547572626F20417373656D626C656C[144]

TGGAAGGGCTAATTCACTCCCAACGAAGACAAGA[145]

143. These difficulties are discussed in the section above, "The Unreliability of Software."

144. These are the first two lines of the hex listing for a modified *Trident Polymorphic Engine*, used to link into an otherwise monomorphic virus. The linked engine will cause the virus to mutate unpredictably and, depending on the virus search tools employed, unrecognizably [194].

145. This commences the nucleotide sequence for the HIV retroviral provirus, responsible for the disease AIDS and associated with adult T-cell leukemia lymphoma [195]. Even relatively early, in 1985 the investigators recognized the apparent polymorphism of the virus [196]. We now know that HIV mutates "incredibly rapidly, often within the very person it infects; there have been instances of viruses in the same individual varying by as much as 30 percent" [197].

A third problem associated with the Registry system is ensuring that a host system can identify certified agents. If AIs are polymorphic, and indeed if we would expect their behavior will be *increasingly* polymorphic, will it be possible to identify these agents sufficiently to allow Turing certification?

The image that comes to mind is nailing Jell-O to the wall. There appears to be no guarantee that the specific code selected for certification will be involved in a pathological episode. Programs can replicate and send portions of themselves to distributed sites for processing; even localized programs can and will modify themselves. Thus, we have the problem of *identifying* a "certified" agent. What does it mean to suggest that "Program Alef," or indeed any constituent agent, has been certified? How will we recognize Alef in a week or in two years? Will we recognize Alef in a small codelet operating at a remote site? The problem arises because copies of digital information are perfectly indistinguishable from the original. An inspection of code or data will not always reveal whether it has been truncated and therefore could have come from a different or uncertified agent. The difficulty arises because Turing certifications apply to only a *portion* of the processing environment: only to polymorphic agents.[146]

It is essential to trace the behavior of a "processing environment" if any form of Turing certification is to succeed. Instead of looking to mutating processing nodes and executable files, though, we should look to the communications channels *between* such nodes. The emphasis remains on

146. The problem of digital polymorphism underlies much of the current anxiety in intellectual property circles [124,125]. Copyright problems arise as programs, at various levels of abstraction, are sampled and appropriated by third parties. See, e.g., [126–128]. These problems are magnified when Internet access allows essentially invisible hypertext linking of widely distributed documents, which eviscerates the integrity of previously discrete documents, texts, and graphics. See generally [129]. Similar issues arise in the context of trademarks in the digital context. The essential distinctiveness of trademarks and trade dress is difficult to fix when morphing software subtly alters texts and design, the touchstones of trademark law, along a spectrum of infinite detail. And trademark law's utter reliance on the notion of "customer confusion" is subtly eroded when customers shop and buy products online, instead of physically inspecting the products in the physical world. For the basics of trademark law, see [130].

Digital sampling disbands that which previously had been a unity. Sampling erodes the underlying notion of property—a notion derived from bounded, tangible physical land—as the logical focus of the law [124]. It remains unclear which notion, or underlying legal metaphor, will replace this type of property. This is not the place to detail these problems, but the reference here does confirm the futility of treating software as identifiable, localized, discrete property. Emerging problems in the digital copyright realm concomitantly suggest problems in the present context. Certifying software, as we might tag wild animals on the veldt for ecological analysis, will not work.

the overall processing environment, but focuses on a direct examination of information flow and *not* on the nodes from which the flow originates or ends. Messages among processing nodes would be "tagged" with the appropriate authenticating codes, and Turing agencies would certify only programs that require the perpetuation of an electronic tag on all messages. Under the proposed model, no message is accepted or processed without the electronic tag. That tag—in effect the encrypted Turing certification—acts as a unique digital thumbprint, a coded stand-in for the program that assures the host processing environment of the bona fides of the messenger data (or instruction).

The model I propose is based on the behavior of viruses.[147] Mutating viruses will change small portions of their RNA (or DNA) sequence, but retain a persisting, distinct sequence otherwise. The small genetic change is enough to produce important differences in viral appearance and behavior, enough, for example, to avoid the effects of a host's immune defenses searching for the former viral incarnation. So, too, certain structures can be locked into digital code without conflicting with the behavior of polymorphic programs. To borrow language from genetics, we can assure ourselves of some genetic persistence even where the phenotype is untraceably polymorphic. Intelligent programs, as such, may be sufficiently dispersed and mutable as to defy firm identification; however, the messages passed by the programs and their codelets can be inspected for certification. Ultimately it is those messages, that information flow, which must be subject to Turing oversight.

The image, then, is of many gatekeepers watching over the flow of traffic. Only on presentation of an authentic "pass" are messages admitted to remote sites, or the value of variables passed to other codelets, subroutines, or peripherals. The gatekeepers do not care where the message came from or whether it is affiliated in some fashion with a Program Alef (whatever that is), as long as it is certified. The overhead of such a system might be high, but that is irrelevant.[148] The point here is simply to model a target for Turing certification that does not depend on permanently fixing the boundaries of an artificial intelligence (or other program). And that can be accomplished. Turing Registries offer the possibility of a technological response to the quandary first posed by technology.

---

147. See the section above "Agents and Networked Viruses."

148. The overhead cost may not in fact be prohibitive or otherwise technically difficult. After a draft of this chapter was completed, a report appeared on the new programming language Java, discussed, for example, at [94]. In language reminiscent of my requirements for Turing certification of processing environments, Java constantly monitors foreign programs for viruses and other rogue programs, both while the foreign modules or applets (i.e., small programs or applications) are resident and after they have been terminated [198].

PERSPECTIVES

> *There is no reason to suppose machines have any limitations not shared by man.*
>
> —Marvin L. Minsky [199]

> *It was a machine's dilemma, this inability to distinguish between actual and programmed experience.*
>
> —James Luceno [200]

Effective, distributed AI presents the problem of judgment. As we rely on our electronic doppelgangers, we increasingly lose the referents of history, honest memory, and truth as measured by correspondence with the way the world really is. Simulation, including computerized simulation, carries us away from these into what Baudrillard calls the "catastrophe of reality" [201]. We forget that digital space follows rules different from those of physical reality; we forget that digital space has no dimension and is ungoverned by the rules of physics, that it is just data [202]. The logic of the simulacrum does not necessarily correlate with the physics of the real world. Simulacra may be simply self-referential, in the way that television shows just quote other shows and movies, and news stories cover the making of a film or "expose" the truth about another news show. A multimedia hyperlink between President Kennedy's death and the CIA does not mean that the two are in fact related. The logic of program flow does not explain how physical objects interact, and digital models may not accurately mimic physical reality.[149]

Describing the 1989 missile attack on the U.S.S. *Stark*, author Gary Chapman recalls that the ship's defense systems probably failed because onboard computers classified the French-made—but Iraqi-launched—missiles as *friendly*. The programmers' assumptions did not correspond with the reality of the Persian Gulf conflict. Chapman writes:

> [W]hen we talk about what goes on in a computer, we're talking about an entire complex of relations, assumptions, actions, intentions, design, error, too, as well as the results, and so on. A computer is a device that allows us to put cognitive models into operational form. But cognitive models are fictions, artificial constructs that correspond more or less to what happens in the world. When they don't, though, the results range from the absurd to the tragic [204].

---

149. See generally [36,62,203].

This disassociation from the real suggests that computer programs may at once be faithful to a digital logic and fail to exercise ordinary judgment. The problem is not new to those who develop artificial intelligence programs. Some researchers assemble enormous databases of facts and lessons [27]; others draft heuristics, rules of thumb, in an effort to provide a reservoir of "common sense" on which programs may draw. Such efforts are probably doomed: they cast us back to the dark days of the rule-bound, domain-bound expert systems briefly described earlier.[150] In any event, these are enormous undertakings, and cannot hope to underpin the distributed intelligences that we may use in the next decade. Other researchers may seek to impose "global controls": constraints on the judgments AIs may make, limits on the data they may consider, and bars to the analogies they may devise. These are *all* attempts to substitute human judgment, formed in advance and so without the key facts, for the judgment of the AI created to handle the situation. AIs again confront us with the old, old issue of how much autonomy to allow the machine.

AIs' potential lack of common-sense sensitivity to the constraints of the physical world presents a serious risk to operations entrusted to autonomous artificial intelligences. But doubtless we will use intelligent systems and pay the price, just as we use automobiles and pay the terrible toll of some 29,000 highway deaths a year [206]. Users and site operators can be circumspect and should carefully screen agents, just as care is needed on the road. But AIs will inevitably arrive at "very strange answers"; when they do, the courts will not provide an effective forum.

As they assess the liability for AIs gone awry, courts will be tempted to use the same sort of proximate cause analysis they always have. But it will not work. We may know that AIs are involved as one of an infinite number of causes in fact. But against the background of ephemeral, distributed, polymorphic processing elements, judges will not be able to pluck out specific program applets, or human agencies, as proximate causes.

There is a risk when courts fail to provide a meaningful remedy for the felt insults of technology. Laws may be enacted and cases decided that make technological development too much of a litigation risk. Where social policy cannot find its target, liability may be imposed on the undeserving, and others may unjustly escape. The Turing Registry may provide a technological option to resolve the legal conundrum.

150. See generally [205].

References

[1] Hofstadter, Douglas, and Melanie Mitchell, *Fluid Concepts and Creative Analogies*, New York: Basic Books, 1995, p. 236.

[2] Baudrillard, Jean, *The Transparency of Evil,* 1993, p. 53.

[3] Munakata, Toshinori, "Commercial and Industrial AI," *Comm. ACM* 23 (March 1994).

[4] Gold, Lawrence, "If AI Ran the Zoo," *Byte* 79 (Dec. 1995).

[5] Dietrich, Eric (ed.), *Thinking Computers and Virtual Persons*, San Diego, CA: Academic Press, 1994.

[6] Dennett, Daniel C., *Consciousness Explained*, Boston: Little Brown, 1991, pp. 431–440.

[7] Searle, John R., *The Rediscovery of the Mind*, 1992.

[8] Weber, Bruce, "A Mean Chess-Playing Computer Tears at the Meaning of Thought," *NewYork Times* pp. A1, A11 (Feb. 19, 1996).

[9] Zuboff, Shoshana, *In the Age of the Smart Machine*, New York: Basic Books, 1988.

[10] Gleick, James, *Chaos: Making a New Science*, 1987.

[11] "The Computer Industry," *The Economist* 20 (Sept. 17, 1994).

[12] Weiser, Mark, "Some Computer Science Issues in Ubiquitous Computing," *Comm. ACM* 75 (July 1993).

[13] Gleick, James, "Watch This Space," *New York Times Magazine* 14 (July 9, 1995).

[14] *The Economist* 4, 10 (Sept. 17, 1994).

[15] "World PC Surge Seen," *New York Times* 39 (July 3, 1995).

[16] "Twentieth Anniversary Report," *Byte* 74 (Sept. 1995).

[17] Prosise, Jeff, "Windows 95 Secrets," 14 *PC Magazine* 247 (Dec. 19, 1995).

[18] Kling, Rob, et al., "Massively Parallel Computing and Information Capitalism," in *A New Era in Computation*, N. Metropolis and Gian-Carlo Rota (eds.), Cambridge, MA: MIT Press, 1993, p. 191.

[19] Caudill, Maureen, and Charles Butler, "The Characteristics of Learning Systems," in *Naturally Intelligent Systems,* 1990, pp. 152, 153.

[20] Orfali, Robert, et al., "Intergalactic Client/Server Computing," *Byte* 108 (April 1995).

[21] "Regulating Cyberspace," 268 *Science* 628 (1995).

[22] "The Computer Industry," *The Economist* 16 (Sept. 17, 1994).

[23] Richter, Jane, "Distributing Data," *Byte* 139 (June 1994).

[24] "Computerworld," *Client/Server J.* (Aug. 1995).

[25] Knowles, Anne, "InterAp Assigns Intelligent Agents to the Web," *PC Week* 42, 47 (June 12, 1995).

[26] Wingfield, Nick, "Internet Apps to Get Intelligent Search Agents," *Infoworld* 16 (May 15, 1995).

[27] Lenat, Douglas B., "CYC: A Large-Scale Investment in Knowledge Infrastructure," *Comm. ACM* 33 (Nov. 1995).

[28] Chaib-draa, B., "Industrial Applications of Distributed AI," *Comm. ACM* 49, 50 (Nov. 1995).

[29] El-Najdawi, M. K., and Anthony C. Stylianou, "Expert Support Systems: Integrating AI Technologies," *Comm. ACM* 55, 56 (Dec. 1993).

[30] King, Julia, and Thomas Hoffman, "Hotel Heading for 'Net Without Reservations," *Computerworld* 1 (Nov. 27, 1995).

[31] King, Julia, and Thomas Hoffman, "Hotel Heading for 'Net Without Reservations," *Computerworld* 28 (Nov. 27, 1995).

[32] Weinberg, Neal, "Computer/phone Finds Its Voice," *Computerworld* 55 (Dec. 18, 1995).

[33] "Cool Today, Hot Tomorrow," *Byte* 84 (Sept. 1995).
[34] "Collision!" *Byte* 199 (Sept., 1995).
[35] "All-in-one Computers: Computer-TV Hybrids Invade the Den," *Byte* 32 (Sept. 1995).
[35] Roberts, Sarah L., "The Internet Comes to Cable," *PC Magazine* 31 (Dec. 5, 1995).
[36] Wiener, Lauren Ruth, *Digital Woes*, Reading, MA: Addison-Wesley 1993.
[37] Reinhart, Andy, "The Network With Smarts," *Byte* 51 (Oct. 1994).
[38] Caudill, Maureen, and Charles Butler, "The Characteristics of Learning Systems," in *Naturally Intelligent Systems,* Cambridge, MA: MIT Press, 1990, p. 25.
[39] Munakata, Toshinori, "Commercial and Industrial AI," *Comm. ACM* 23–25 (March 1994).
[40] Rumelhart, David, et al., "The Basic Ideas in Neural Networks," *Comm. ACM* 87 (Mar. 1994).
[41] Rumelhart, David, et al., "The Basic Ideas in Neural Networks," *Comm. ACM* 89 (Mar. 1994).
[42] Machrone, Bill, "Care and Feeding of Neural Nets," *PC Week* 63 (June 5, 1995).
[43] Haruhiko Asada, "Representation and Learning of Nonlinear Compliance Using Neural Nets," 9 *IEEE Transactions on Robotics & Automation* 863 (Dec. 1993).
[44] May, Gary S., "Manufacturing ICs the Neural Way," *IEEE Spectrum* 47 (Sept. 1994).
[45] Caudill, Maureen, and Charles Butler, "The Characteristics of Learning Systems," in *Naturally Intelligent Systems,* Cambridge, MA: MIT Press, 1990, pp. 241–60.
[46] "Technology Focus," *Computer Design* 74 (Sept. 1994).
[47] Hofstadter, Douglas, and Melanie Mitchell, *Fluid Concepts and Creative Analogies,* New York: Basic Books, 1995, p. 97.
[48] Hofstadter, Douglas, and Melanie Mitchell, *Fluid Concepts and Creative Analogies,* New York: Basic Books, 1995, p. 124–25.
[49] Caudill, Maureen, and Charles Butler, "The Characteristics of Learning Systems," in *Naturally Intelligent Systems,* Cambridge, MA: MIT Press, 1990, p. 155.
[50] Hofstadter, Douglas, and Melanie Mitchell, *Fluid Concepts and Creative Analogies,* New York: Basic Books, 1995, p. 256.
[51] Hofstadter, Douglas, and Melanie Mitchell, *Fluid Concepts and Creative Analogies,* New York: Basic Books, 1995, p. 91.
[52] Hinton, G., et al., "The 'Wake-Sleep' Algorithm for Unsupervised Neural Networks," 268 *Science* 1158 (1995).
[53] Minsky, Marvin, *The Society of Mind*, New York: SImon & Schuster, 1985.
[54] "A Conversation with Marvin Minsky About Agents," *Comm. ACM* 23, (July 1994).
[55] Hofstadter, Douglas, and Melanie Mitchell, *Fluid Concepts and Creative Analogies,* New York: Basic Books, 1995, p. 215.
[56] Hofstadter, Douglas, and Melanie Mitchell, *Fluid Concepts and Creative Analogies,* New York: Basic Books, 1995, p. 235.
[57] Ludwig, Mark A., *Computers, Viruses, Artificial Life and Evolution,* Tuscon, AZ: American Eagel Publications, 1993, p. 86.
[58] Sheridan, Thomas B., *Telerobotics, Automation, and Human Supervisory Control,* Cambridge, MA: MIT Press, 1992, pp. 356–60.
[59] Vadlamudi, Pardhu, "Corel Faces Buggy Software Backlash," *InfoWorld* 25 (Oct. 23, 1995).
[60] Littlewood, Bev, and Lorenzo Strigini, "The Risks of Software," *Sci. Am.* 62 (Nov. 1992).
[61] Blum, Manuel, and Sampath Kannan, "Designing Programs That Check Their Work," 42 *J. Ass'n Computing Mach.* 267 (1995).

[62] Littlewood, Bev, and Lorenzo Strigini, "Validation of Ultrahigh Dependability for Software-based Systems," *Comm. ACM* 69 (Nov. 1993).
[63] Littlewood, Bev, and Lorenzo Strigini, "Validation of Ultrahigh Dependability for Software-based Systems," *Comm. ACM* 78–79 (Nov. 1993).
[64] Wiener, Lauren Ruth, *Digital Woes*, Reading, MA: Addison Wesley 1993, pp. 96–98.
[65] Wiener, Lauren Ruth, *Digital Woes*, Reading, MA: Addison-Wesley 1993, p. 96.
[66] Davey, Tom, "Chip Makers Split on Listing Bugs," *PC Week* 115 (Dec. 18, 1995).
[67] "Editorial," *Byte* 126 (Sept. 1995).
[68] Joch, Alan, "How Software Doesn't Work," *Byte* 49–50 (Dec. 1995).
[69] "Intelligent Agents," *Comm. ACM* (July 1994).
[70] "Software Agents Prepare to Sift the Riches of Cyberspace," 265 *Science* 882 (1994).
[71] *Intelligent Agents: Theories, Architectures and Languages,* 1995.
[72] Masaru, Deiri, "Knowbotics: A Talk on the Wild Side with Pattie Maes," *InterCommunications* 110 (Annual 1994).
[73] BIS Strategic Decisions, *Pragmatic Application of Information Agents*, 1995.
[74] Reinhardt, Andy, "The Network With Smarts," *Byte* 51 (Oct. 1994).
[75] Halfhill, Tom R., and Andy Reinhardt, "Just Like Magic?" *Byte* 22 (Feb. 1994).
[76] Fitzgerald, Michael, "Agent Technology Stirs Hope of Magical Future," *Computerworld* 37 (Jan. 31, 1994).
[77] Lee, Yvonne L., "Telescript Eases Cross-network Communication," *Infoworld* 22 (Jan. 17, 1994).
[78] "Agents of Change," *Byte* 95 (Mar. 1995).
[79] Wayner, Peter, "Agents Away," *Byte* 113 (May 1994).
[80] Durfee, Edmund, and Jeffery Rosenschein, "Distributed Problem Solving and Multi-Agent Systems: Comparisons and Examples," in *Distributed Artificial Intelligence, Proceedings of the 13th International Distributed Artificial Intelligence Workshop,* published by the American Association for Artificial Intelligence, 1994, p. 52.
[81] Edmonds, Ernest, et al., "Support for Collaborative Design: Agents and Emergence," *Comm. ACM* 41 (July 1994).
[82] "Software Agents Prepare to Sift the Riches of Cyberspace," 265 *Science* 883 (1994).
[83] Masaru, Deiri, "Knowbotics: A Talk on the Wild Side with Pattie Maes," *InterCommunications* 114–115 (Annual 1994).
[84] Liu, JyiShane, and Katia Sycara, "Distributed Problem Solving Through Coordination in a Society of Agents," in *Distributed Artificial Intelligence, Proceedings of the 13th International Distributed Artificial Intelligence Workshop,* published by the American Association for Artificial Intelligence, 1994, p. 169.
[85] Huberman, Bernardo A., "Towards A Social Mind," 2 *Stan. Human. Rev.* 103, 107 (1992).
[86] Huberman, Bernardo A., and Tad Hogg, "Distributed Computation as an Economic System," *J. Econ. Persp.* 141 (Winter 1995).
[87] Hayes-Roth, Frederick, and Neil Jacobstein, "The State of Knowledge-Based Systems," *Comm. ACM* 27, 35 (Mar. 1994).
[88] Hayes-Roth, Frederick, and Neil Jacobstein, "The State of Knowledge-Based Systems," *Comm. ACM* 38 (Mar. 1994).
[89] Markoff, John, "Staking Claim in Alternative Software on the Internet," *New York Times* p. D4 (Sept. 25, 1995).
[90] Venditto, Gus, "Java: It's Hot, But Is It Ready To Serve?" *Internet World* 76, 78 (Feb. 1996).
[91] December, John, *Presenting Java*, 1995, p. 6.
[92] Ritchey, Tim, *Java!* 1996, pp. 15–16.
[93] Ritchey, Tim, *Java!* 1996, pp. 23–24.

[94] "Beyond Java: Distributed Objects on Web," *PC Week* 48 (Dec. 18, 1995).
[95] Johnston, Stuart J., and Kim S. Nash, "Capitulation!" *Computerworld* 1 (Dec. 11, 1995).
[96] van Hoff, Arthur, et al., *Hooked on Java*, 1996, p. 17.
[97] van Hoff, Arthur, et al., *Hooked on Java*, 1996, p. 17–18.
[98] Ritchey, Tim, *Java!* 1996, pp. 50–51, 99.
[99] "Netscape Flaw Could Cause Harm," *Marin Indep. J.* p. C3 (Feb. 22, 1996).
[100] Anthes, Gary, "Still a Few Chinks in Java's Armor," *Computerworld* (Feb. 19, 1996).
[101] Reinhardt, Andy, "The Network With Smarts," *Byte* 51, 64 (Oct. 1994).
[102] Wayner, Peter, "Agents Away," *Byte* 116 (May 1994).
[103] "Security Flaws Force Microsoft to Get Active," *ComputerWorld* (Feb. 24, 1997).
[104] "Housebroken Virus," *InfoSecurity News* (Sept./Oct. 1994).
[105] Hoffman, Lance J. (ed.), *Rogue Programs: Viruses, Worms and Trojan Horses*, 1990.
[106] Ludwig, Mark A., *The Little Black Book of Computer Viruses*, Tuscon, AZ: American Eagle Publications, 1991.
[107] Louderback, Jim, "A Virus by Another Name Causes Equal Pain," *PC Week* 106 (Apr. 3, 1995).
[108] Pontin, Jason, "Macro Virus Threat Continues," *InfoWorld* 36 (Dec. 11, 1995).
[109] Anthes, Gary, "Macro Viruses Pose Hazard to PC Health," *Computerworld* 45 (Feb. 19, 1996).
[110] Vizard, Michael, "Security Woes Dull OLE Luster," *Computerworld* 1 (Dec. 6, 1993).
[111] Huberman, Bernardo A., "Towards A Social Mind," 2 *Stan. Human. Rev.* 107 (1992).
[112] "In Short," *Info. Wk.* 88 (May 22, 1995).
[113] "Win95 Wizard Leads to Confusion," *InfoWorld* 6 (May 29, 1995).
[114] Tribe, Laurence H., "Rights of Privacy and Personhood," in *American Constitutional Law,* Minneapolis: West Publishers, § 15, 1988.
[115] "The Web-Crawler Wars," 269 *Science* 1355 (1995).
[116] Frentzen, Jeff, "Spiders, Worms, and Robots Crawl in the Web," *PC Week* (Feb. 13, 1995).
[117] Holland, John, *Adaptation in Natural and Artificial Systems*, Cambridge, MA: MIT Press, 1992.
[118] Holland, John, *Adaptation in Natural and Artificial Systems*, Cambridge, MA: MIT Press, 1992, pp. 3–4.
[119] Livingston, Brian, "How Microsoft Disables Rivals' Internet Software," *InfoWorld* 42 (Sept. 25, 1995).
[120] Raymond, Eric (ed.), *The New Hacker's Dictionary*, Cambridge, MA: MIT Press, 1991, p. 134.
[121] Gelernter, David, *Mirror Worlds*, New York: Oxford University Press, 1992, p. 54.
[122] Lee, Atkinson, and Mark Atkinson, *Using Borland C++*, Carmel, IN: Que, 1991, p. 432.
[123] Comaford, Christine, "Inside the Black Box of Objects," *PC Week* 18 (Nov. 20, 1995).
[124] Barlow, John Perry, "The Economy of Ideas," *Wired* 84 (Mar. 1994).
[125] Samuelson, Pamela, "The NII Intellectual Property Report," 37 *Comm. ACM* 21 (Dec. 1994).
[126] Nimmer, Raymond T., et al., "Software Copyright: Sliding Scales and Abstract Expression," 32 *Houston L. Rev.* 317 (1995).
[127] Karnow, Curtis, "Data Morphing: Ownership, Copyright, and Creation," in this volume.
[128] Samuleson, Pamela, "Digital Media and the Law," 34 *Comm. ACM* 23 (Oct. 1994).
[129] Carter, Mary E., *Electronic Highway Robbery: An Artist's Guide to Copyrights in the Digital Era*, Berkeley, CA: Peach Pit Press, 1996.

[130] McCarthy, J. Thomas, *McCarthy on Trademarks and Unfair Competition*, 1995.
[131] Lewis, C. S., *The Screwtape Letters*, p. xxi.
[132] Witkin, Bernard, *Summary of California Law*, 5, § 3, 1988.
[133] Harper, Fowler V., et al., *The Law of Torts*, 3, § 18.8, 1986.
[134] Harper, Fowler V., et al., *The Law of Torts*, 4, § 14.3, 1986.
[135] Harper, Fowler V., et al., *The Law of Torts*, 4, § 20.2, 1986.
[136] *Restatement (Second) Of Torts*, § 433 B (3), 1965.
[137] Harper, Fowler V., et al., *The Law of Torts*, 4, § 20.4, 1986.
[138] *Restatement (Second) of Torts*, § 431 cmt. a, 1965.
[139] Prosser, William L., *The Law of Torts*, 270, 1971.
[140] Harper, Fowler V., et al., *The Law of Torts*, 4, § 20.5, 1986.
[141] Lewyn, Mark, "Eavesdroppers Speak Softly and Carry a Big Scrambler," *Business Week* 6 (Oct. 3, 1994).
[142] James, "The Qualities of the Reasonable Man in Negligence Cases," 16 *Mo. L. Rev.* 1, 5–15 (1951).
[143] Winograd, Terry, and Fernando Flores, *Understanding Computers and Cognition*, 1986, p. 155.
[144] Sheridan, Thomas B., *Telerobotics, Automation, and Human Supervisory Control*, Cambridge, MA: MIT Press, 1992, p. 341.
[145] Winograd, Terry, and Fernando Flores, *Understanding Computers and Cognition*, 1986, pp. 155–56.
[146] Hayes-Roth, Frederick, and Neil Jacobstein, "The State of Knowledge-Based Systems," *Comm. ACM* 27, 38 (Mar. 1994).
[147] Maes, Pattie, "Artificial Life Meets Entertainment: Lifelike Autonomous Agents," *Comm. ACM* 108 (Nov. 1995).
[148] Maes, Pattie, "Artificial Life Meets Entertainment: Lifelike Autonomous Agents," *Comm. ACM* 111 (Nov. 1995).
[149] "Security Schemes Aspire to No-Fuss System Protection," 270 *Science* 1113–1114 (Nov. 1995).
[150] Krol, Ed, *The Whole Internet User's Guide & Catalog*, Sebastopol, CA: O'Reilly & Associates, 1992, pp. 47–48.
[151] Durfee, Edmund, and Jeffery Rosenschein, "Distributed Problem Solving and Multi-Agent Systems: Comparisons and Examples," in *Distributed Artificial Intelligence, Proceedings of the 13th International Distributed Artificial Intelligence Workshop*, published by the American Association for Artificial Intelligence, 1994, p. 54.
[152] Durfee, Edmund, and Jeffery Rosenschein, "Distributed Problem Solving and Multi-Agent Systems: Comparisons and Examples," in *Distributed Artificial Intelligence, Proceedings of the 13th International Distributed Artificial Intelligence Workshop*, published by the American Association for Artificial Intelligence, 1994, p. vii.
[153] Huhns, Michael, et al., "Global Information Management via Local Autonomous Agents," in *Distributed Artificial Intelligence, Proceedings of the 13th International Distributed Artificial Intelligence Workshop*, published by the American Association for Artificial Intelligence, 1994, p. 120.
[154] Sikora, Riyaz, and Michael J. Shaw, "Manufacturing Information Coordination and System Integration by a Multi-Agent Framework," in *Distributed Artificial Intelligence, Proceedings of the 13th International Distributed Artificial Intelligence Workshop*, published by the American Association for Artificial Intelligence, 1994, p. 314.
[155] Parunak, H. Van Dyke, "Deploying Autonomous Agents on the Shop Floor: A Preliminary Report," in *Distributed Artificial Intelligence, Proceedings of the 13th Inter-*

national Distributed Artificial Intelligence Workshop, published by the American Association for Artificial Intelligence, 1994, p. 259.

[156] Rao, Anand S., and Graerne Murray, "Multi-Agent Mental-State Recognition and its Application to Air-Combat Modeling," in *Distributed Artificial Intelligence, Proceedings of the 13th International Distributed Artificial Intelligence Workshop,* published by the American Association for Artificial Intelligence, 1994, p. 264.

[157] Weihmayer, Robert, and Hugo Velthuijsen, "Application of Distributed AI and Cooperative Problem Solving to Telecommunications," in *Distributed Artificial Intelligence, Proceedings of the 13th International Distributed Artificial Intelligence Workshop,* published by the American Association for Artificial Intelligence, 1994, p. 353.

[158] Sheridan, Thomas B., *Telerobotics, Automation, and Human Supervisory Control,* Cambridge, MA: MIT Press, 1992, pp. 239–45.

[159] Taylor, Richard, *Instrument Flying,* 1978.

[160] Jeppeson Sanderson, *Advanced Pilot Manual,* 1981.

[161] Jörg P., Müller, and Markus Pischel, "Integrating Agent Interaction into a Planner-Reactor Architecture," in *Distributed Artificial Intelligence, Proceedings of the 13th International Distributed Artificial Intelligence Workshop,* published by the American Association for Artificial Intelligence, 1994, p. 232.

[162] Finin, T., et al., "KQML—A Language and Protocol for Knowledge and Information Exchange," in *Distributed Artificial Intelligence, Proceedings of the 13th International Distributed Artificial Intelligence Workshop,* published by the American Association for Artificial Intelligence, 1994, p. 99.

[163] Huhns, Michael, et al., "Global Information Management via Local Autonomous Agents," in *Distributed Artificial Intelligence, Proceedings of the 13th International Distributed Artificial Intelligence Workshop,* published by the American Association for Artificial Intelligence, 1994, p. 128.

[164] Hofstadter, Douglas, and Melanie Mitchell, *Fluid Concepts and Creative Analogies,* New York: Basic Books, 1995, p. 235.

[165] Sheridan, Thomas B., *Telerobotics, Automation, and Human Supervisory Control,* Cambridge, MA: MIT Press, 1992, pp. 356–60.

[166] Prosser and Keeton, *The Law of Torts,* § 43, 1984, pp. 293, 297-99.

[167] Harper, Fowler V., et al., *The Law of Torts,* 4, § 20.5, 1988.

[168] *Restatement (Second) of Torts,* § 281 cmts. e & f, 435(2), 451, 1965.

[169] *Restatement (Second) of Torts,* § 281 cmts. e & f, 450–451, 1965.

[170] Harper, Fowler V., et al., *The Law of Torts,* 4, § 20.5, 1986, p. 162.

[171] Hardy, I. Trotter, "The Proper Legal Regime For 'Cyberspace,'" 55 *U. Pitt. L. Rev.* 993, 1013, 1040 (1994).

[172] Crothers, Brooke, and Bob Fracis, "Flawed IDE Controller Corrupts Data," *InfoWorld* 6 (Aug. 14, 1995).

[173] Rheinfrank, John, "A Conversation with John Seely Brown," 11 *Interactions* 43, 50 (1995)

[174] Karnow, Curtis, "The Encrypted Self: Fleshing Out the Rights of Electronic Personalities," in this volume.

[175] Neumann, Peter, "Distributed Systems Have Distributed Risks," 39 *Communications of the ACM* 130 (November 1996).

[176] Baudrillard, Jean, *The Illusion of the End,* Stanford: Stanford University Press, 1994, p. 40.

[177] Gibson, William, *Neuromancer,* New York: Ace Science Fiction, 1984.

[178] Harel, David, *The Science of Computing,* Reading, MA: Addison-Wesley, 1987, p. 202.

[179] Churchland, Paul, et al., "Could A Machine Think?" in *Thinking Computers & Virtual Persons*, Eric Dietrich (ed.), San Diego, CA: Academic Press, 1994, pp. 157–158.

[180] Sheridan, Thomas B., *Telerobotics, Automation, and Human Supervisory Control*, Cambridge, MA: MIT Press, 1992, p. 348–49.

[181] Zimmermann, Philip, *The Official PGP User's Guide*, Cambridge, MA: MIT Press, 1995.

[182] Schireson, Max, "Decoding the Complexities of Cryptography," *PC Week* 84 (Jan. 10, 1994).

[183] Schneier, Bruce, *Applied Cryptography*, 1994.

[184] Barlow, John Perry, "A Plain Text on Crypto Policy," 36 *Comm. ACM* 11, 21 (Nov. 1993).

[185] Levy, Steven, "Crypto Rebels," *Wired* 54 (May-June 1993).

[186] Froomkin, A. Michael, "The Metaphor Is the Key: Cryptography, the Clipper Chip, and the Constitution," 143 *U. Pa. L. Rev.* 709 (1995).

[187] Erickson, Jonathan, "Cryptography Fires Up the Feds," *Dr. Dobb's Journal* 6 (Dec. 1993).

[188] Hayes, Brian, "The Electronic Palimpsest," *The Sciences* 10 (Sept.-Oct. 1993).

[189] Karnow, Curtis, "Encryption and Export Laws: The Algorithm as Nuclear Weapon," in this volume.

[190] "Corporations Eye Private Security Systems," *Byte* 36 (Aug. 1995).

[191] Wingfield, Nick, "Digital IDs to Help Secure Internet: Certifying Authorities to Promote Electronic Commerce," *Infoworld* 12 (Oct. 23, 1995).

[192] Corcoran, Elizabeth, "Reston's CyberCash Joins Internet Companies' Rush to Wall St.," *The Washington Post* p. C1 (Dec. 23, 1995).

[193] Keeton, Robert E., *Insurance Law Basic Text*, 1971, p. 318.

[194] Ludwig, Mark A., *Computers, Viruses, Artificial Life, and Evolution*, Tuscon, AZ: American Eagle Publications, 1993, p. 356.

[195] Ratner, Lee, et al., "Complete Nucleotide Sequence of the AIDS Virus, HTLV-III," 313 *Nature* 277 (1985).

[196] Ratner, Lee, et al., "Complete Nucleotide Sequence of the AIDS Virus, HTLV-III," 313 *Nature* 283 (1985).

[197] Radetsky, Peter, *The Invisible Invaders*, 1991, p. 341.

[198] Markoff, John, "A Software Language to Put You in the Picture," *New York Times* p. C1 (Sept. 25, 1995).

[199] Minsky, Marvin L., *Computation: Finite and Infinite Machines*, 1967, p. vii.

[200] Luceno, James, *The Big Empty*, New York: Ballentine Books, 1993, p. 82.

[201] Baudrillard, Jean, *The Illusion of the End*, Stanford: Stanford University Press, 1994, p. 113.

[202] Couchot, Edmond, "Between the Real and the Virtual," *InterCommunications* 16 (Annual 1994).

[203] Karnow, Curtis, "Information Loss and Implicit Error in Complex Modeling Machines," in this volume.

[204] Chapman, Gary, "Making Sense out of Nonsense: Rescuing Reality from Virtual Reality," in *Culture on the Brink: Ideologies of Technology*, Gretchen Bender and Timothy Druckrey (eds.), Seattle: Bay Press, 1994, pp. 149, 151.

[205] Hofstadter, Douglas, and Melanie Mitchell, *Fluid Concepts and Creative Analogies*, New York: Basic Books, 1995, p. 319.

[206] "Highway Safety—Causes of Injury in Automobile Crashes," GAO Rept. No. GAO/PEMD-95-4, May 9, 1995, p. 1.

Part VI
Crime and Punishment

Many hackers—myself included—have stepped over the arbitrary line that the government has drawn separating legal from illegal. But that's not a requirement of hacking. It just happens to be a common side effect. For most hackers, the intent is not to vandalize, break laws, or terrorize. It's to learn and explore. ...Hacking is embedded in the history and culture of computers and communications.

—Joel Snyder, *Internet World*

Criminal law is "public law" in the sense that its scope and requirements, originally enacted by public bodies such as legislatures, cannot be modified by private agreement. Private agreement can have a profound effect on other area of the law, such as contract and torts. For example, private agreement can shift the liability for a loss or for someone's negligence and can even insulate a person entirely from being sued for a wide variety of acts.

But private parties cannot do much to stop the relentless effects of the criminal law. The acts of legislatures can be reviewed to see if they are constitutional, and prosecutors can in their discretion decide not to indict crimes—but otherwise persons in the physical jurisdiction of a governmental entity can do nothing but try to avoid breaking the law. And there are many governmental entities. In the United States, we are subject to the federal law, which exerts its predominating influence over all means of interstate commerce, such as the use of the highways, aircraft, and the telephones lines that carry facsimile, conversations, and computer communications. And then each of the 50 states maintains a full panoply of criminal laws, as do the local governments of cites and counties.

The jurisdiction of each of these entities, though, is not limited to persons found physically within their borders. A jurisdiction's laws usually apply to persons *outside* the area if that person's actions have a bad *effect* inside the interested jurisdiction or (sometimes) if the crime is committed in some *part* within the borders of the jurisdiction. So, for example, a crook in New York taking money from innocents in Alaska may be indicted in Alaska. A client of mine in London was indicted in Los Angeles because he assertedly assisted Los Angeles taxpayers in hiding money from the Internal Revenue Service; when the FBI tricked him into flying into the United States, the arrest was valid. States can prosecute receivers of stolen goods as they drive through, just as federal grand juries can indict anyone who does any part of a crime in their district. And when an attorney general of a certain northern state stated that he felt free to indict anyone who, from anywhere in the United States, caused illegal emails to flow through his state, it was a threat on which he conceivably could have made good [1]. This is a troubling proposition, though, because (1) we have no idea what route our emails will take; indeed, a single message may be divided up, be routed via a dozen paths, and recombined at the destination,[1] and (2) sending an email can constitute an enormous variety of crimes, from blackmail, extortion, terrorist threats, and threats against the President, to criminal obscenity, stalking, hate crimes, and criminal trade secret offenses,[2] as well as constituting civil offenses such as sexual harassment and libel. Any many, many offenses are either defined differently in divers jurisdictions, or simply do not exist in other jurisdictions. Gambling may be legal in Las Vegas, on the Mississippi River, and on various Indian reservations, but gambling is not legal in every location where one can log into a gambling web site [3]. The crime of obscenity is expressly dependent on local community standards, but photographs on an electronic bulletin board that garner a repulsed shrug in San Francisco will send someone to jail for two to three years in Memphis, Tennessee—even if the defendant and the computer holding the pictures are in the Bay Area and the defendant never sets foot in Tennessee.[3]

Some states have truly unpredictable regulations that probably apply to citizens in other states. For example, Maryland law makes it illegal to distribute (or possess with intent to distribute) a sound or image recording whose packaging does not identify the name and address of the performer. There are doubtless innumerable web sites that carry audio or bits of film clips that violate this law. And Georgia makes it a crime to set up a web page that uses marks, logos, and other trademark indicia which might be seen as making a false representation that the site is endorsed by

1. See, e.g., [2].

2. See, e.g., the Federal Economic Espionage Act of 1996 (15-plus years imprisonment), 18 U.S.C. § 1831, 1832.

3. *United States v. Thomas*, 96 CDOS 609 (6th Cir. January 29, 1996).

the mark owner. This local law implicitly alludes to federal trademark law, but has no necessary dependence on it. As Chapter 12 reveals, many routine actions undertaken in the context of networked computers appear to break laws, contrary to the expectations of many users.

It is fair to suggest, then, that important difficulties are posed by the combination of (1) potentially nationwide jurisdiction for (2) a variety of overlapping criminal laws, (3) the effect of which, in some circumstances, can neither be foreseen nor avoided by preemptive agreements and contracts. The next chapter explores some of these issues in the application of public criminal law in the context of networked computers.

References

[1] Rosenoer, Jonathan, "The Long Arm of the Law," *Wired* 109 (April 1997).

[2] "Regulating Cyberspace," 268 *Science* 628 (May 5, 1995).

[3] Betts, M., and G. Anthes, "Online Boundaries Unclear: Internet Tramples Legal Jurisdictions," 29 *Computerworld* 1, 16 (June 5, 1995).

Recombinant Culture: Crime in the Digital Network

12

Technically, I didn't commit a crime. All I did was destroy data. I didn't steal anything.

—Martin Sprouse, from the case of a Bank of America employee who planted a logic bomb in the computer system [1]

A floating world of liquid media where the body is daily downloaded into the floating world of the net, where data is the real, and where high technology can finally fulfill its destiny of an out-of-body experience....In recombinant culture, the electronically mediated body comes alive as our android other....

—Arthur Kroker [2]

Reality has left the physical world and moved into the virtual one.

—Benjamin Woolley [3]

INTRODUCTION

The signal shift in the development of the digital culture is the loss of physical laws as the conclusive arbiter of action.

Physical and natural laws have generally governed what was possible, and so they provided limits to everything from politics to property. Hence the notion of a "reality check." This grounding in physical reality provided a certain minimum commonality among human experiences, providing the basis for shared assumptions. As an intractable limit, the test of physical reality sorted out the real from the dream, the valid from the invalid, the true from the false, the effective from the futile, and even the good from the bad.

Correspondingly, the legal system has always been a system of ascertainable limits, with physical "property" as it is central core and metaphor. When intangible things of value became valuable, the system bred "intellectual property" to cover these intangibles. The legal system rapidly extended its reach to all manner of vaporous, indiscernible stuffs—we own, and can steal, and can sue over: invisible trade secrets, accounts receivable, expectations of profit, and patented ideas. By and large these intangibles, and all of the intangible personal rights protected by constitutions and law, plainly do find tangible expression or location. The patent is written out on paper, the right of privacy inheres in our physical selves. The intangible stuff is actually parceled out on a per capita basis, and documents are kept that certify the individual owners and in so doing mark out the boundaries, the borders, among these incorporeal things. That's why we can reasonably call the stuff "intellectual property."

But we are leaving the physical world behind, and with it the touchstone of physical and natural laws, together with the notion of irreducible limits. Increasingly, the things we deal with are located in a digitized networked space [4,5]. This is marked by the remarkable increase in computing power available to the average user[1] and growth of the networks.[2] The social work of the culture increasingly is mediated by the computer. Too, the productivity of the culture—the making of valuable things—is done on computers, and those valuable things are often bit-composites stored on those machines.

The stuff in computers is information. Some people want to use information in ways that horrify others. Disputants call on the legal system

1. We see about 50% to 70% more computing power per year, and memory density doubles every 12 to 18 months. Between 1978 and 1994, desktop transistor density went up 100 times, and raw computing power *over 500 times*:

	80x86 Performance-Max MIPS	Transistor density
June '78	.75	29,000
1982	2.66	134,000
April '89	70	1,200,000
March '93	112	3,100,000
March '94	166	3,300,000

2. Note the size and growth of the Internet: 21,000 connected networks, 60 countries, 15 million users; 2 million computers, with a rate of *monthly* growth of 7% to 10%. See also [7].

to settle the matter, and some people get indicted. The law has to provide answers, but there is no consensus on what the rules are: the technology is growing too fast, and there is too much myth and ignorance. Instead of social consensus, the federal and state governments have passed laws, hundreds of them, criminalizing so-called unauthorized access and data transmission. Other groups, too, have called for laws, regulations, bills of rights, cyberspace constitutions, and so on, to regulate the electronic arena.

These efforts are futile.

UNDERSTANDING INFORMATION

There is a deep confusion on the value of data and information.

Those who value it most treat it with no honor. They may take a perverse pride in its mutation, in its endless potential, in its infinite mutability. The apostles of the information age tell us that data is free [8], that the wineskins have burst [9], that information belongs to us all. Of course, information that is as ubiquitous as air, cheap as sand, is barely property at all. These same people believe profoundly in the value of privacy and are the vanguard of those combating the government's efforts to control the keys to private encryption. They oppose the Clipper chip, federal government monopolies on encryption devices, corporate control over personal and credit information, and other threats to the individual's power to exercise total domain over the data that he creates. They see—correctly, I think—that control over and access to information, on the one hand, and control over the citizenry, on the other hand, are inseparable.

On the other side, those charged with the enforcement of the law know well that our movement to an information society is transforming notions of property. They think that if digital property cannot be protected, then property as such will be eviscerated, and with it one of the central foundations of the legal system. It's true; tampering with data can cost millions of dollars; some information is private and shouldn't be released; some data storage areas, and entire systems, should be off limits to unauthorized personnel. The digital world, despite its increasing size and universality—or indeed as a result of it—does not stop at the keyboard and electron tube: it reaches out into the physical world and operates subway systems, nuclear power plants, passenger aircraft, and the economies of developed countries. Precisely because much of what we call physical reality is controlled by computers, we care very much about what happens to specific data. Criminal law *must* handle the real-world consequences of these electronic events.

But most judges, juries, and prosecutors don't understand the technology. It's difficult to spend time and money on investigations of system break-ins, it's hard to ask for jail time when the only harm was a temporary slowdown of a network, and it's troublesome to try a case when the perpetrator simply looked around a few files and left a little electronic graffiti. Why lock someone up for peaking over the digital fence? Who's been hurt? What's the loss in value of picked-over data? When the criminal justice system is wholly overwhelmed by traditional crimes of violence, murder, and drugs, then electronic crime takes a back seat. It is too difficult to understand and too complicated to spend time on [10].

There is a deep ambivalence over the value of information.

PARCELING OUT INFORMATION AS PROPERTY

The law and the culture it symbolizes may not comprehend the nature of data and information, but the law does know what property is. For this reason developments in computer law[3] consist by and large of (1) developing the definition of "property" in the digitized world,[4] and (2) the adumbration of proscribed actions done with that newly broadened definition. Laws address the interception of data, accessing data and systems, tampering with and altering data, obtaining data of value, frustrating authorized users from accessing data, and viewing other's use of data, including keystroke monitoring (18 U.S.C. § 2510; [Federal] Computer Fraud and Abuse Act).[5]

---

3. Every state but Vermont has some form of computer-specific criminal statute [11]. See generally [12].

4. For example, Massachusetts found it necessary to pass a law stating that "property" includes "electronically processed or stored data, either tangible or intangible, [and] data while in transit...." Mass. Gen. Law Ch. 266, § 30(2)(1992). In its efforts to combat computer crime, Missouri (like many other states) now defines property to include "information," any electronic data, and indeed any intangible item of value, Missouri Stats. § 569.093 (10). Kansas defines property similarly for the same purpose, Kan. Crim. Code § 21-3755 (i)(h). California makes it a crime to, without permission, use or copy computer data, or to access that data, or to prevent others from accessing it, Cal. Penal Code § 502(c)(2), (4), (5) and (7). New York's law, typically, also focuses on unauthorized access to and tampering of computer data, N.Y. Penal Law §§ 156.10; 156.20.

5. "Currently 26 million employees are monitored at work, and this number should increase as computers become more widely used and the cost of monitoring systems decreases. If employers monitor the number of keystrokes we make a minute to assess our productivity, or read our electronic mail, what will they be allowed to do in the future as the human-machine interface becomes even more personal?" [13].

Computer crime laws first appropriate the entire digital world, claiming the right to control the movement of every electron and fiber cable photon; then these laws carve out the boundaries of criminal behavior through the notion of *unauthorized access*. Exquisite borders circumscribe each grouping of data; criminal behavior is the breach of the border.

The coverage of these laws can be stunningly broad: every computer and every attached device and all communication facilities "related to or operating in conjunction with" the system can be protected under federal law (18 USC § 1030(e)), and in Pennsylvania it can be a crime to interrupt the "normal functioning" of a system or organization, whatever that means.[6]

For example, federal law focuses directly on the unauthorized access of a computing system, and refines that by also making it a crime *to exceed* one's authorized access to secure, for example, financial information.[7] A 10-year jail term can be imposed for preventing others from their authorized use of a system.[8] It is a federal felony to have an unauthorized access to a system for the purpose of any copying or taking or disclosing or possession of anything of value.[9] Any access that exceeds one's authority and which thus secures or alters any information, or which prevents an authorized user from having access, is a federal crime.[10]

Thus, computer crime laws follow the general legal presumptions of discrete, separable areas surrounding and defining each legally cognizably entity: homes and property around individuals protected by trespass laws; a web of trade secrets and confidentiality around corporations; and privacy interests marking them all off from each other. These are the traditional rules of consanguinity, and the digital universe has been subject to the same colonization. Every user's access is protected from everyone else's. Every digital interference, from destroying another's data to degrading the computer system (Del. Code Ann. tit. 11 § 934),[11] can be a crime.

Law defines the borders, and so the very existence and shape, of legal entities. When the breaches are mild, use the civil law and sue; when the breaches are considered severe, enter the criminal law.

---

6. Pa. Cons. Stats. tit. 18 § 3933.

7. 18 USC § 1030.

8. 18 USC § 1030(a)(5).

9. *Id.*

10. 18 USC § 2701.

11. Many states proscribe denying or interrupting computer services to authorized users. For example, Pa. Cons. Stat. tit. 18 § 3933 (a)(1); Cal. Penal Code § 502 (c)(5).

FAULTY TRANSITION TO THE NEW NETWORK

Run, run, run as fast as you can
You can't catch me—
I'm the Gingerbread Man.

These efforts, are, like much of the law's attempts to play catch up with technology, too little, and too late. The legal system, inherently conservative, is perpetually a decade behind the technology[12] and so the governing law can rarely be directly invoked. As I note later, though, the *price* for this divergence is paid by technology as it finds itself shackled by unfit law.

There are two central reasons for the law's difficulty in treating electronic material as property. First, data is infinitely mutable and so untraceable; second is the nature of the network.

Infinitely Mutable Data

The character of this chaos is set by the technology's apparent disregard for the laws of physics, its chameleon skin and infinite capability for morphing. The outstanding fact of data, and of the algorithms that operate on them, is the property of *recombinancy.* Originals are in no way distinguishable from copies; the very notion of an "original" is empty. The same data can be manifested as sound, text, and image, and thus each as an algorithmic equivalence that defeats conventional notions of copying and reference. Every transformation is of an equivalent validity. Transmogrification is now banal. Continuity, accuracy, and truth are a function, thus, of the algorithm; these become a purely mathematical concept. The integrity of the data substrate is wholly divorced from the infinite colors and shapes and forms of its manifestations and appearance.

We cannot tell where the data came from, what the data is, who made it, when it was made, or if it mutated, or was stripped of identifying information such as trademark or copyright claims. Only the computer knows, and its memory is volatile. It has permanent amnesia.

---

12. An appellate opinion read today took four years to get through the trial system, and about another year and a half in the court of appeal. Of course, the technology had been out on the consumer market before the case was initially filed and so was a few years old then. The court of appeal decided the meaning and effect of a statute that dated back to the filing of the case, some six or seven years ago, and which had actually been passed a year or so before that, in response to the technology of the day. Now, one knows what a nine-year-old statute means. (And the statute has probably since been amended.)

Digital Networks

The truth is that we do not use separable computers. We no longer access separable information and data. We do not really have discrete data in separate banks of RAM or a discrete CPU in every garage. The slow, painful efforts of the legal system to accommodate the computer are mooted by the advent of the network.

Lawmakers see communications devices and other connections simply as appendages to the key individual computer and data vaults. The image is of a network as simply the medium by which the data travels and by which criminals access property. The image here is of streets connecting homes, the medium of air between communicating humans, empty space between the stars.

But that image is false.

For the network reveals that there are no borders, and there never will be. The ultimate networked machine is wholly wired, operating across time and space at the speed of light, connecting every user to every database. This ideal is a sort of single parallel processing multithreaded machine, and it isn't far from reality. The fact of interconnectedness is as critical to the nature of the computing as the connected nodes themselves. Just as electricity itself has no meaning absent the conducting wires, so too the evolving computing environment is *essentially* a connected unitary whole.[13]

This essential fusion produces a deep quandary for the new computer crime codes. There are problems with the obscenity law of Tennessee affecting bulletin board system (BBS) operators in Arkansas and California; there are problems with the United States trying to restrict the so-called export of encryption software across the physical U.S. borders while a billion transfers of the same software across the 50 states are legal. There are issues with applying classic telephone wiretap law to the interception of email within a company. It is not clear if an unauthorized access has occurred when an authorized user inadvertently introduces into a system a virus written by someone without access. It is not possible to determine if a repair technician with approved "supervisor" access to a network can legally look at data. We don't really know if privacy rights protect files on local hard drives, and whether different rules apply to hard drives across the state or national borders. We know the U.S. government can't insist on the right to record keystrokes at my stand-alone PC,

13. "[C]yberspace...[is] the place where our lives and fates are increasingly determined.... The power of this realm comes from its connectedness. It is a continuum, not a series of discreet systems that act independently of each other. Blips are not isolated events" [14].

but the FBI wants the power to record and read keystrokes if sent by wire.[14]

This is utter confusion. There are different laws with different results regulating a single activity. Federal and state law conflict on privacy rights and criminal conduct; we have conflicts between the laws of two or more states, all of which simultaneously apply to the same networked environment.

To catch up with technology, states have extended the definition of "property" to electronic impulses:[15] but in the electroverse of infinitely mutable replicant data, that net sweeps far too wide. Under this brave new regime, theft can include logging onto the wrong partition, a felony erupts when an unwanted comment appears on a startup screen,[16] and you could get jail time for playing the networkable multiplayer games DOOM or Falcon on your employer's network.[17] Think of it this way: DOOM is a sort of Trojan horse, a virus *pretending* to be a game.

Indeed, there are a host of unexpected behaviors we've seen explode from the products of commercial vendors that could qualify as felonies under the new laws. For example: *compression* software that slows down a system, operating systems that trash files, telecommunication devices that lose mail, or any iota thereof. There is a bug—I mean a feature—in the Windows tool OLE 2.0 which allows one to send a hidden program inside a document, defeating many routine security techniques.[18] Sounds like unauthorized access to me.

---

14. Which is why fiber-optic cable and its bland transmission of a thousand pulses of light makes some people in the government ill. (Optical fiber is *not* what George Bush meant when he referred to a thousand points of light.) For how can the FBI "tap" photons?

15. See, i.e., Montana Code § 4-5-2-310, which covers any input to or any output from any computer or network.

16. Conn. Gen. Stat. § 53a-251 (e); Del. Code Ann. tit. 11 § 935. See also [15].

17. Prosecutors have almost unlimited discretion to indict anyone they want. They will go ahead and do just that when the statute says that theft of property is a crime, and then defines property as electronic cycles; or if the law criminalizes any unauthorized use of system; or makes it a crime to "degrade" a system: that's exactly what DOOM will do. That game program uses broadcast messages, up to 100 per second, across the network, which degrades the system and can even crash it. See [16]. Multiuser dungeons (MUDs), too, have been banned from universities and entire continents (Australia) because of their degrading impact on systems resources [17].

18. The problem is one that could affect distributed objects generally. In the case of OLE 2.0, the embedded item appears to be data, but when activated (for an edit, for example) causes the current program to yield control, including its menu bar, to the parent program that created the data; that parent program could be anywhere on a network, an unknown quantity to the present user. See, e.g., [18]. As a related concern, it is practically impossible to keep track of the copyright issues governing objects shared among applications, or swapped in from remote sites. See [19].

But *criminal* unauthorized access? Crooks, surely, must "intend" to do evil before they are locked away. A noted programmer describes how to make a simple modification to a C compiler that will be undetectable and cause a miscompilation under set circumstances. "If this were not deliberate, it would be called a compiler 'bug.' Since it is deliberate, it should be called a Trojan horse" [20]. But the "deliberate-mistake" dichotomy is not quite how the criminal law operates. "Reckless disregard" of the consequences, knowing enough so that one *ought* to know the consequences, is usually enough for a conviction. As one my favorite jury instructions notes, we are all presumed to know and *intend* the ordinary consequences of our actions—and that can be a lot of consequences. Presumably someone at Microsoft knew of the OLE bug; presumably Borland knew when it released C++ 4.0 that it generated a rounding error [21]. Both these bugs, of course, in effect destroy data; arguably, Microsoft and Borland "intended" that destruction. And the "deliberate-mistake" dichotomy does not correlate with the risk posed by viruses, either: some like the Macintosh Scores virus (and perhaps, too, the famous Internet worm) probably was released *unintentionally* into the general population [22]. In both cases, though, many prosecutors would indict and secure a conviction under the new computer crime statutes.

Is that too bizarre? A prosecution for a bug that eats data or that provides an entry point for data modification? Not bizarre at all. Decisions to indict are essentially unreviewable, and often made by prosecutors with a weak grip on the technology. The same can be said for the juries that decide on guilt.

INTERLUDE: STATE OF THE NET

Let us pause to probe deeper into the nature of networked computing. There are two central aspects that bear scrutiny. Those aspects are: (1) agents and related objects, what I term *self-directed code*, and (2) the worldwide electronic environment in which those agents operate: webbed databases.

Let me briefly describe these.

Self-directed code means those programs that can be let loose to find their way and report back with some sort of result. These exist in rudimentary form now, in everything from Borland's and Microsoft's spreadsheets to General Magic's draft telecommunication programs and Java's applets. These "agent" objects combine code and data, are independent, sensitive to their context, and therefore do not need complete instructions from the human operator. In that sense, these agents are self directing: they don't need direct human supervision. The point is to embody some amount of artificial intelligence and thus be able to execute some unpredicted computer tasks.[19]

The second salient aspect here is the development of worldwide hyperlinked databases, utterly transparent to the user.[20] Already, users of the Internet's World Wide Web are familiar with the invisible shifting of data sources. True global networking is the ultimate, realizable, ambition.[21] Every network now accepts and operates with the notion of a *logical drive*, as opposed to the physical drive resident in determined physical space. The simple notion of logical drive, logical space as oppose to physical space, makes it plain that *a network "exists" everywhere that any user can access.* Noting the physical location of our data is utterly pointless.[22]

Imagine, now: self-directing code in a worldwide habitat of hyperlinked data.[23] That is our model network.

This electroverse is a concatenation of endlessly looping data and symbols that do nothing ultimately but refer to themselves and to combinations of themselves. Here, the physical and natural worlds are leveled, electrocuted, no more than a series of signs and symbols.

This is the nebulotic data soup. It is an unrestrained, frenzied hyperbole of text, sound, and graphics, each moment a cut and paste morphed version of others, an endless processing and transmission of the bit-

19. See generally [23].

20. Nearly three-quarters of *Fortune* 100 companies spread their data across multiple databases. The number was expected to rise to 86% by 1996 [24].

21. For a list of products devoted to global networking, see [25].

22. Operating systems now all incorporate this sort of transparency: that's part and parcel of the "networkable" approach to such systems. See [26], regarding the distributed computing environment IBM planned for 1995, including virtual directory trees for enterprise-wide nets, email with agents. Microsoft's Windows 95 and Windows 97 operating systems—which unlike Windows 3.1, are true operating systems including a replacement for DOS—introduce "applications built around networks and a built-in capacity to share information. E-mail, for example, will be accessible from the tool bar" [27]. Microsoft's Windows 97 and Netscape's Consellation complete this transition to the integration of the desktop with the World Wide Web [28]. Finally, for an example of a patent describing data displays independent of the physical hardware, see "Multiple Virtual Screens on an 'X-Window' Terminal," U.S. Patent No. 5,289,574 (issued February 23, 1994 [Hewlett Packard, assignee]).

23. "The seamless nature of object systems will radically alter the way we think about *where* our data is. Data will be encapsulated in objects that will in some cases be able to roam to where they are most needed. We are in the habit of thinking that a document is simply stored on a particular hard disk. Distributed object systems will ask us to surrender that comfortable certainty in exchange for the power and flexibility of location-transparent storage" [29].

stream. Memory is a looping self-replicating tape: there is no past, or there is an infinity of pasts.

Here we have the infinite geography of the electronic cosmos operating at lightspeed: communication so fast and transparent that the elements, the actors, the agents of communication are swept up into the transmission stream and lose all identity but for their existence as transmission agents, each a repeating station, each no more than input and output. Each one is a copper wire linked into other wires, until we have a single endlessly looping strand, truly *e pluribus unum.* The electroverse is zero culture, inhabited by android shadow-selves who fear no law, abide no punishment, and feel no guilt.

It is in this digital soup, this is a hyperrelational environment, which we see as the death of the barrier. We have no cells, we have no inside and outside, we have no public world, and we have no private world. What we do have is the network and the death of dichotomy. This is fatal for the legal system, which depends for its very life on the existence of barriers—after all, that's what the law does: it utters the line between *this* and *that,* and punishes the transgressor.

But our android shadows cannot be punished.

CRIME IN A PHANTASMAGORICAL TERRAIN

> *There are probably no secure systems on the Internet.*
>
> —Peter Neumann, SRI International, quoted in [30].

What does this pose, this universe of warpspeed propagation of data and signs, this self-metastasis, this cancer of replicating code? The problem is one of extreme difficulty and consequence, because while the law *evaporates* as the network is perfected, the *need* for law grows as the network expands to control the infrastructure of the real world. The self-contained replicant synthesis of networked data actually resides in the physical world; it incorporates real human beings who live and eat and die. The truth is that the digital world truly affects and is affected by the physical world: it operates nuclear power plants, subways systems, and passenger aircraft. "[A] blip in the money markets can raise bank lending rates, a blip in a multinational's productivity can close factories and throw economies into depression, a blip in the TV ratings can wipe out an entire genre of programming, a blip in an early warning system can release a missile" [14]. The essential connectivity of the network generates at once its extraordinary power, its volatile reactivity, and so its striking vulnerability.

This, then, is the nub of the problem. As I have noted, the solution to date has been to pass more and more laws, more regulations, calls for new constitutions of cyberspace, to invoke new rights, new privileges; to heap laws on laws. The confusions regarding data and information that I described earlier have been addressed in just this way by all sectors of the political spectrum: The FBI wants more laws,[24] liberal Harvard law school professor Lawrence Tribe wants more laws,[25] the National Security Agency wants regulation of encryption, employers want rules on employee privacy rights, online services such as Genie and Prodigy regulate "inappropriate" and "offensive" speech,[26] Congress is considering new copyright law to make it perfectly clear that electronic copying *is* copying.[27] This is the all-American way: more legislation. Don't let anything escape; take no prisoners.

This impetus is fueled by the cynical acceptance by the government—the legislatures and law enforcement—that electronic reality really does exist as a sort of parallel universe. They have come to accept the proposition that virtual reality—what I have termed tongue-in-cheek the electroverse—has an existence independent, legally speaking, of the day-to-day reality governed by traditional law. Denizens of that place call it cyberspace, a high frontier, a new territory precisely like the Moon or some other undiscovered country.

The lawmakers have endorsed that schizophrenia pressed by the electro-cognoscenti, and they have endorsed it out of fear, ignorance, and misunderstanding. How else to react to the omniscient, omnipotent power of cyberspace? Well, take its unruly tenants at their word (they seem to know what they are talking about), and treat cyberspace as a competing reality: regulate it and break it up into chunks called property.

But alternative universes provide a very bad model: Neither law nor technology benefits. The law founders and sinks in the clear blue fungible sea of the network. And the electronic community is on the verge of being legislated to death, ruled out of fear and loathing, chained by broad and

24. See, e.g., the telephony bills introduced by the government, which would ensure the ability to conduct court-approved wire taps.

25. Tribe proposed a new electronic bill of rights in his keynote address to the first Computer, Freedom & Privacy conference (CFP).

26. See, e.g., [31], regarding forums shutting down because the subject matter might be inappropriate for young children; messages with religious, ethnic, and other references being deleted; and users being warned of possible termination.

27. Congress was asked to consider a series of amendments to the Copyright Act, which among things would state that an electronic transmission is a "publication" of a protected work, that copyright holders have electronic distribution rights to their works, and so on. See "Intellectual Property Working Group Draft Report," released July 7, 1994. Telnet iitf.doc.gov (gopher login) /speech testimony & documents.

detailed laws that can make anything—the movement of an electron—illegal.

The notion of cyber law—legal rules peculiar to the electronic communications context—makes no more sense than print law, newspaper law, movie law, TV law, shopping center law, video game law, or, indeed, washing machine law. The network is *not*, in fact, a place like the Moon. It is a tool, a machine, like a tractor or a pen or Velcro, an extension of the physical, moral human being who in turn is subject to the mundane legal system. We can have no law of *place* or property here, and so new computer crime law's heavy regulation of data and authorized access is quite wrong-headed. Computer crime is not really about unauthorized access,[28] the actions of self-directed code in the network are not "intended" in any sense recognized by the criminal law, and it can be impossible to tell if anyone "owns" any given chunk of infinitely mutable data.

The battle of the metaphor always erupts in the face of new and powerful technologies. Our imagination is fired, but our stability is threatened, and we always seek precedent for understanding. So we use the *property* analogy; the metaphor of invaded homes and goods when systems are attacked, the allusion to space and universes. But this is a category mistake. Computer-mediated "space" is no more space than DNA is a person, no more than digital signals are a picture or a novel. Bits and bytes are not the equivalent of their manifestations; the genotype is not the phenotype.

The criminal law has no business here. For the network has no borders, and the autonomous space of hyperperfect illusion and flawlessly recombinant culture is too slippery for any statute.

But, should then the wily hacker deploy with impunity? Shall we let a thousand viruses bloom? What *is* the reasonable role of the law in view of the fact that networked computers operate our transportation, banking, power plants, military, and other key infrastructures?[29]

Control—the control we need—is not finally a *legal* problem at all. It is a social, moral, and technological problem. The law simply will not save us from the next full stealth polymorphic virus; but widely accepted social norms and technological shields might. Back up the data, use firewalls to insulate machines and data, use smart cards and arbitrary, chang-

---

28. Ironically, and most importantly, nearly 80% of "computer criminals" are actually "insiders with verified system access" [32]. The FBI has reported that in 80% of its computer crime investigations, the Internet was used to gain illegal access to systems [33]. But unless Ms. Stewart's figures are wrong, the FBI's estimates are more a function of the FBI's interests than a reflection of the real world.

29. "Today, with almost every detail of modern life controlled or influenced by computers and communications-driven systems, our infrastructure has an exposed underbelly: software" [34].

ing passwords; employ private cancelbots; use real encryption for authentication, privacy, and safe commercial transactions. Never make the mistake of believing that a *computerized* system is necessarily an improvement, and think about not using computers, at all, for certain very high-risk tasks.[30]

When real people suffer real injury, measured as real financial loss, then indict and convict those who demonstrably intended that harm. That's the purview of the criminal code. If a jury of ordinary people, using ordinary laws on fraud and theft, wouldn't convict, then leave it alone. Put the armed criminals in cells and leave the android shadows to their electric dreams. There are grey areas, to be sure, but there *should* be grey areas, governed slowly by developing manners and custom, not crushed out by an omnipresent criminal code.

References

[1] Sprouse, Martin (ed.), *Sabotage in the American Workplace: Anecdotes of Dissatisfaction, Mischief and Revenge,* 1992.
[2] Kroker, Arthur, *Spasm*, New York, 1993, p. 36.
[3] Woolley, Benjamin, *Virtual Worlds*, Cambridge, MA: Blackwell, 1992, p. 235.
[4] Zuboff, S., *In the Age of the Smart Machine*, New York: Basic Books, 1988.
[5] Woolley, Benjamin *Virtual Worlds*, Cambridge, MA: Blackwell, 1992, pp. 133–134.
[6] "80x86 Evolution," 19 *Byte* 88 (June 1994).
[7] "Special Report: Distributed Computing," 19 *Byte* 125 (June 1994).
[8] Wiener, Lauren, *Digital Woes,* Reading, MA: Addison-Wesley, 1993.
[9] Barlow, John Perry, "The Economy of Ideas," 2.03 *Wired* 84 (March 1994).
[10] Gemignani, Michael, "Viruses and Criminal Law," in *Rogue Programs: Viruses, Worms, and Trojan Horses,* Lance Hoffman (ed.), New York, 1990, p. 99.
[11] "Computer-Related Crimes," Note, 30 *American Criminal Law Review* 495, 513n.144 (1993).
[12] Branscombe, Anne, "Rogue Computer Programs and Computer Rogues," in *Rogue Programs,* p. 59.
[13] Pickover, Clifford, *Visions of the Future: Art, Technology, and Computing in the 21st Century,* 2d ed., St. Martin's Press, Introduction.
[14] Woolley, Benjamin *Virtual Worlds*, Cambridge, MA: Blackwell, 1992, p. 133.
[15] Branscombe, Anne, "Rogue Computer Programs and Computer Rogues," in *Rogue Programs,* p. 68.
[16] "Games Storm LANs," 5 *Infosecurity News* 8 (July/Aug. 1994).
[17] Kelly, Kevin, et al., "The Dragon Ate My Homework," 1.3 *Wired* 69, 73 (July/Aug 1993).
[18] Coffee, Peter, "Distributed Objects Form Info Highway Hazards," *PC Week* 80 (April 18, 1994).
[19] Brandel, William, "Objects Spur User's Licensing Concerns," 28 *Computerworld* 1 (July 4, 1994).

---

30. A number of writers, including this author at the 1993 Las Vegas DEFCON conference, have suggested dispensing with computers for certain tasks [8,35,36].

[20] Thompson, Ken, "Reflections on Trusting Trust," in *Rogue Programs,* pp. 121, 125.
[21] Coffee, Peter, "Close enough isn't good enough in computer math," *PC Week* (Feb. 14, 1994).
[22] Stefanac, Suzanne, "Mad Macs," in *Rogue Programs,* pp. 180, 189.
[23] "Intelligent Agents," 37 *Communications of the ACM* (July 1994).
[24] "Fractured Data Reported," 5 *Infosecurity News* 9 (July/Aug. 1994).
[25] "The Best of Interop+Networld," *Byte* 36 (July 1994).
[26] Polilli, Steve, et al., "IBM will launch revamped E-mail strategy in fall," 16 *InfoWorld* 1, 75 (July 4, 1994).
[27] Coursey, David, "The Death of the Single User," *PC World* 53 (June 1994).
[28] "Web-Centric Windows 97," *PC Magazine* 28–29 (March 4, 1997).
[29] Wayner, Peter, "Objects on the March," *Byte* 139 (Jan. 1994).
[30] Lewis, Peter H., "Hackers on internet pose security risks," *The New York Times* p. C19 (July 21, 1994).
[31] Lewis, Peter H., "Censors Become a Force on Cyberspace Frontier," *The New York Times* p. 1, col. 1 (June 29, 1994).
[32] Stewart, Nina, 4 *InfoSecurity News* 32 (May/June 1994).
[33] Anthes, Gary, "Internet panel finds reusable passwords a threat," *ComputerWorld* 28 (March 28, 1994).
[34] Black, Peter, "Soft Kill," 1.3 *Wired* 49–50 (July/Aug. 1993).
[35] Littlewood, Bev, et al., "The Risk of Software," 267 *Scientific American* 62 (Nov. 1992).
[36] Karnow, Curtis, "Legal Implications of Complex Virtual Reality Systems," DEFCON, 1, July 1993, available at ftp.fc.net/pub/defcon/DEFCON-1/text/karnow-1.doc

Bibliography

Baudrillard, Jean, *The Transparency of Evil,* London, 1990.

Barlow, John Perry, "The Economy of Ideas," 2.03 *Wired* 84 (March 1994).

Ludwig, Mark, *The Little Black Book of Computer Viruses,* Tucson, AZ: American Eagle Publications, 1991.

The Algorithm As Nuclear Weapon: Encryption and Export Laws

13

The United States government lists strong encryption software together with nuclear weapons on its Munitions List, barred from export to other countries without a formal State Department license. While the software can be transmitted across state lines and used by any U.S. resident without official concern, a violation of the export ban can result in a one-million-dollar fine and up to 10 years in a federal penitentiary.[1]

The law, known as the Arms Export Control Act (AECA), is the subject of much concern among those interested in electronic privacy and civil rights. The law also poses serious problems for software developers interested in competing in the international marketplace; for them, this note provides a basic outline of the law and the associated practical issues.

Encryption engines scramble and unscramble data, ensuring privacy and security. Some products, such as those based on public key encryption, also authenticate the data: the recipient of the message can be assured that the message was in fact encrypted by a certain person, creating a "digital signature" that guarantees identity across anonymous electronic networks. (These mechanisms are described in works listed at the end of this chapter)

Encryption is essential. Corporations are extending their reach across national lines, databases are ramifying across physical boundaries, and large-scale networks are erupting as the core technology of business, government, and academia throughout the globe. Computers as stand-alone processing stations are a thing of the past: Nets own the future. Corporate assets are now digital, stored and accessed electronically around

1. 22 USC § 2778(c). This chapter was written before certain procedural changes went into effect in 1996–1997, including changes in the federal agencies nominally in charge of export control. The points made in this chapter, however, remain valid.

the world. But with the spread of the Net, only encryption can protect these assets from the prying eyes of competitors, the media, and curious or hostile governments. Thus we see the powerful demand for strong encryption in the basic applications: word processing, databases, and network and telecommunications software. Even operating systems may now include the technology.

But the AECA normally blocks the export of products with this indispensable technology. The rationale, of course, is that the country's enemies would otherwise secure a "weapon" with which to defeat our surveillance capabilities. But roughly 200 products from 22 countries—including our "allies" Great Britain, France, and Japan—use encryption currently barred from export. At least 58 foreign software packages use the data encryption standard (DES).

Not all encryption products are blocked: very weak encryption (such as the easily broken data shuffling engine that comes with most word processors) is not, but the Munitions List and the governing regulations, the International Traffic in Arms Regulations (ITAR), officially known as the Defense Trade Regulations, are very broad. These cover unclassified information, including everything from blueprints for military vehicles to scientific papers on algorithms for designing fault-tolerant logic circuits, the detection of signals in noise, and developments in metallurgy. Widespread encryption algorithms, such as the almost-20-year-old DES and common freeware programs that use the RSA public key encryption standard, are banned from export as "auxiliary military equipment." And the notion of an "export" is so broad it covers a *domestic* technology disclosure to foreign nationals, including participation in U.S. briefings and symposia.[2]

Compliance with such a broad and ambiguous law is, to put it mildly, very difficult. While core violations are easily condemned (e.g., secretly shipping missiles to Iran) it is unclear whether a court would uphold a conviction at the periphery, such as lecturing on publicly known encryption algorithms at a U.S. seminar attended by foreign graduate students.

The ITAR at § 120.10(5) exempts technical data in the "public domain" but it is not known at this writing whether the State Department will honor the exception and allow, for example, the export of software now widely available via the Internet and formally placed in the public domain. (At this writing a test case was on appeal in the State Department.)

But there is plainly a serious risk of prosecution should a developer export an application with built-in encryption, and an excitable prosecu-

2. 22 C.F.R. §§ 120.10 (d); 125.03 (1983).

tor might indict a software developer simply for using a foreign national on a development team within the United States.

To ensure compliance with the law, one should first review the regulations found at 22 C.F.R § 121.1, which lay out the controlled technologies. (These regulations, like all federal regulations, can be found at any good public library.) Without an export license, the wise developer will take pains to separate foreign nationals from any work involving the restricted technology.

There is a complex web of procedures and regulations governing the required registration of manufacturers and exporters of the listed technologies, and various departments defer to each other depending on the specific technology; suffice it here to note that such registration is necessary. In addition, a license application must be sent into the State Department's Office of Defense Trade Controls (DTC) for every contemplated export. An initial informal contact with the DTC is advisable: the political situation, specific technology at stake, and the contemplated destination may all affect the type and amount of information required by the DTC. Informal contacts can be initiated with a call to the DTC at (703) 875-6644. In reality, the nation's spy agency, the National Security Agency, controls the issuance of these licenses, and perceived national security interests will decide the administration's response.

It may be odd to equate crypto algorithms with F-16 fighters and nuclear-tipped missiles, but that's what the law does. The penalties for noncompliance are severe: developers in the international market need to know the rules.

Bibliography

"Controls On The Export of Militarily Sensitive Technology: National Security Imperative or U.S. Industry Impediment?" 18 *Rutgers Computer & Technology Law Journal* 841 (1992).

"National Security Controls on the Dissemination of Privately Generated Scientific Information," 30 *UCLA L.Rev.* 405 (1982).

"Public Cryptography, Arms Export Controls, and the First Amendment: A Need for Legislation," 17 *Cornell International Law Journal* 197 (1984).

Rubenstein, Ira, "Export Controls and Immigration Law," 93-3 *Immigration Briefings* 1 (March 1993).

"Special Report: Data Security," *IEEE Spectrum* 18–44 (Aug. 1992).

Schneier, B. *Applied Cryptography*, 1994.

Walker, Stephen T., Testimony before the House Subcommittee of Economic Policy, Trade and Environment, Committee on Foreign Affairs, October 12, 1993.

Zimmermann, Phil, *Pretty Good Privacy* (PGP) (Public key encryption freeware, together with two-volume user's manual discussing both the technology and policies associated with encryption), available by FTP to U.S. citizens: (for noncommercial use) ftp net-dist.mit.edu, in the /pub/PGP directory. A FAQ all about PGP is available via FTP at rtfm.mit.edu in /pub/usenet/news.answers/pgp-faq. Documents relating to import/export restrictions, and the procedures involved in securing an export license, are found at an ITAR document site: ftp.eff.org/pub/EFF/Policy/Crypto/ITAR_Export/ITAR_FOIA. See also, http://www.pgp.com

A recent article summarizing some choices for developers is found at Thom Stark, "Encryption for a Small Planet," *Byte* 111 (March 1997).

Part VII
Technology, Society, and the Law

Technology influences us, and is influenced by us, as we adopt and reject its advances, as we deliberate on its implications, as we embrace and reject potential consequences. The Industrial Revolution is an obvious example. The 1950s (and on) were marked by the dissemination of nuclear weapons: we adopted them, rejected them, debated them. Relatively few nuclear weapons were detonated, and few of us ever saw one, but we were in some sense their offspring. We have had similarly intense relationships with the freeway system, the telephone, television, computers, and more particularly the computer network.

Strong new technologies change our language, they introduce new guiding metaphors, and they thereby ultimately change the way in which we think about our culture and ourselves. The legal system is an artifact of the society, no more and no less than drama, literature, film, and the rest of the culture's manifestations: It is pushed, prodded, and mutated as are other cultural expressions. Even lawyers may have a professional interest in cultural expression.

Earlier sections of this book presented chapters on the self, its replacements, substitutions, and digital embellishments; this present section collects a few short chapters on computer technology and the exterior worlds of art, culture, and the *sine qua non* of free cultural expression, the guarantee of free speech established by the First Amendment to the Constitution. In this book's conclusion, "Terminus," strands from the complementary sections on self and context are rewoven and extrapolated.

Where do computers and culture mix? Computing machines have insinuated themselves into music, both as a platform for the making of music and as inspiration and dictator in the substance of the music. Machine sounds, digital models, replicant sampling, looping, and recombinant compositions are found in music, printed, and sculptural art.

"Complex algorithmic equations take flight from mathematics, reappearing in the disguised form of digital sound, images, and smells" [1]. At the same time, as chapters in earlier parts of this book noted, the place of social economic interaction is increasingly the Internet. Museums are on the Net, as are art shows. Web sites are in the permanent collection of the San Francisco Museum of Modern Art. New forms of interactive and collaborative artistic expression are erupting as performers find easy paths to self-publication and to the blended uses of diverse media such as sound and film. The digital experience inspires Hollywood, novels, and fat articles in *The Economist* magazine, and it breeds new multimillion-dollar companies in Silicon Valley. The open sesames of the Internet—those magic letters *www, .com,* and *http://*—now appear on delivery truck and street sign advertising. From high to low culture, from vanguard to the banal, computer technologies are pervasive.

Many of the commentators and authors mentioned in the balance of this book's chapters are convinced that the digital context is now fast becoming all-embracing, all-powerful, all-defining. What is true in cyberspace, it is implicitly asserted, is true generally, because we are all now so deeply influenced by the computer network that we are essentially a function of that technology. Lanham's book, reviewed here, and Monsieur Baudrillard, whom I lavishly cite in "Terminus," are both liable to that generalizing infatuation—as are some of these essays.

This all brings us to an uneasy place, for there is an obvious danger here. Most of the planet does not have Internet access, and even in the Western developed nations the use of computers is often incidental, used merely for adding and typing. But the numbers are always, always increasing: the number of web sites, the number of Internet users, the numbers of employees using computers for communications (as email and intranets expand)—all these numbers are on an ever-accelerating upward slope. Perhaps more pointedly, digital technology is wielded by many of those who make the artifacts of Western culture and by those who popularize those artifacts. Not all are equal in this realm; there is no democracy of artistic or cultural leadership. Finally, it does not matter too much—in this context—that digital technology is not in the hands of all: It will have its effect, anyway. The technology of machines, and their software souls, press on us everywhere.

But as with all technological developments, judgment is required as we select from the menu of futures. It is trite to recall that such developments are double-edged, but enthusiasm can indeed blind us. We know now the bitter lessons and exorbitant price of the industrial revolution; we know what automobiles have done to the environment and to the victims of highway accidents; we know about Chernobyl; and so on. We may be changed irremediably by developing technologies, but we are not helpless before their onslaught. Judgment, based on our recalled past and the past taught to us by history, is an authority that must be brought to bear on the use of technology.

The novel problem presented by computer technology is that it threatens the very basis of that authoritative past. The computer has no memory—no way to distinguish between new and old; between past and present; among the real, the copy, the imitation, and the virtual; or between replicant sampling and an authentic recording. Its advocates have moved past its text-making capabilities to putative postliterate uses; indeed, the Internet did not see its stunning growth until visually based web browsers and concomitant graphical web sites took over from the purely text-based electronic bulletin boards and FTP sites. As Professor Lanham gloats in a book reviewed here, pictures can take over from words; mouse clicks can do what writing accomplished; and meaning (or something like it) can be transmitted with colored fonts, dispensing with words. The technology threatens to eviscerate our past, and with it the only source of our ability to intelligently choose and control the technology's direction.

Computers have a related and more general effect: the destruction of context and with it the notion of meaning. Data inside a machine or sent to the screen via telephone wires comes stripped of its history—or the history can be simply and easily stripped without evidence of the desecration. On the Internet, authoritative information competes with gibberish—but it all look exactly the same. Jokes and serious attacks, which take their sense from context, are often confused. The context of book jackets, covers, the feel of pages—all of the attributes that we use to distinguish types of books—are absent; as are the indicia of wear and tear, of time and history, of an object in a physical world subject to the world's context. A canvas with a hole ripped in it tells us that the item is incomplete, injured; a tattered portion tells us that it was ripped from a larger piece. But there are usually no digital tear marks, which is why there is no essential distinction to be made between a digital original and copy. The stripping of context can make it impossible to know whether one is interacting with a human or a program. Destroying paper card catalogs in favor of online searches murders entire levels of meaning and context, from the old ink handwriting to the contexts generated by the physical proximity of cards, all of which had enlivened intuition. Digital recording, generally, wipes out emendations, scrawls, essays, and evidence of error. That is a great loss, indeed; a loss of the method and means by which our great classics—and most trivial contracts, shopping lists and hasty notes—were created.

When frames and context and history are gone, then meaning is at risk. If all data were transformed at once to a digital format and we all communicated only via digital signals—processor to processor, as it were—we might lose anthropology, archeology, and history—all of it.

Imagine nuclear weapons in everyone's hands, weapons with an odd effect on the brain of the user: touching the item brings immediate knowledge of how to detonate it but eviscerates memory on the effects of a detonation. Computer technology is of like force; like a virus, it carries the seeds of its own expansion and of its host's demise.

♦ ♦ ♦

We hope to glory in our augmented power, without the loss of the culture's literacy that has made us who we are. The power of the processor will inevitably transform the culture, but we can resist its worst incursions if, at the same time, we become truly conservative and protect the past in its accustomed form. We must not mistake the substance for its shadow: original paintings, live opera, tangible sculpture, thick dusty leather books, and the sound of a human voice from a human body—these may keep us safe, if only we will permit them and have them in our home.

The legal system performs as a brake, too; a conservative reminder with explicit appeal to precedent, literacy, the past, and its authoritative nature. We preserve ourselves, in the way meant by this note, every time we invoke the United States Constitution or the Bill of Rights. We resist the transitory when we act on the basis of precedent, as lawyers and judges are meant to do. We preserve context and meaning to the extent that the immediacy of present experience is rejected and, instead, is assimilated by the structure and symbols of the past. The justice system is replete with those structures and symbols, from its marbled halls, the black robes, the incantations and formulas in formal notices, to the bits of Latin that hover, just a bit out of sight, in the names of doctrines and claims. Above all, the law provides a structure for experience: It classifies and categorizes actions and states of mind; it divides expression into admissible evidence and, for example, inadmissible hearsay; it creates and enforces rights, determines the existence of legal entities, determines and segregates lawbreakers from the law abiding, compels the desegregation of the races, and bars offensive sexual advances. The legal system plays off its ambiguous relationship with social morality, and in effect prescribes minimum acceptable social and cultural behavior. The legal system thus acts on the basis of authoritative, highly literate tradition. Judges are lawless in the strongest sense of the term when they ignore precedent and decide instead based on an immediate, personal view of what is right. By contrast, courts that concede the power of the past to judge the present will help us understand that present and to put it into a context.

Other chapters here have emphasized the frustrations of a glacial, backwards-looking legal system on the development of technology. But there is something else to consider, as well: The law, as with other persisting arts that are loath to change, can help us give meaning to the nebulous, lightening-fast technological present.

Reference

[1] Krocker, Arthur, *Spasm,* New York, 1993, p. 39.

14 Review of *Culture, Technology, and Creativity in the Late Twentieth Century*[1]

This volume collects essays from a variety of British authors, directed to the relationship between advanced technology and the arts, primarily the visual arts. Film—and its sound and special visual effects—is a special focus. (In the introduction, editor Hayward explains the goal of the project by citing Leonardo da Vinci and the journal *Leonardo* as precursor commentators on the effect of technology on the arts.)

There is a nice balance here. We have pieces from artists such a recording engineers, directors, musicians, on the one hand; and from the commentators, we have substantial asides on the hyperreal and Baudrillard, a survey of the 1960s Art and Technology movement, and a concluding piece citing self-aware superintelligent global Nets, cyberspace, Mandelbrot, and all the rest. We have pompously absurd items: "the knowledge-seeking subject stands before the paradox of the world—which is to be before all that is in and out of sight, contemplating the terror of the unknown, of the unknowable reality or the terror of the loss of the unreality of real representations." This is found somewhere under the subtitle "The Chronopolitical Nexus" in Tony Fry's contribution, "Art Byting the Dust." And on the other side of cogency, we have pieces on digital technologies in music making (Aland Durant, "A New Day for Music?") that explores the crisis of authorship, copyright, and integrity that digitization brings to the arts, as well as Andy Darley's "From Abstraction to Simulation: Notes on the History of Computer Imaging." This latter item provides both a brief history of recent (since the 1950s) developments in digital film imagery and a concise comment on the role of the entertainment industry in what used to be purely a technical process.

The salient concern here is the perceived battle between content, or art, on the one hand, and the technology on the other. To be sure, the

1. Philip Hayward (ed.), London: John Libbey & Company, received October 1994.

latter is used to make the former, but these essays are infused with a dark doubt. There is a concern here not simply that the use of high technology will be too visible, overshadowing the artistic content (as if the special effects in Star Wars eviscerated notice of the plot, characters, and references to old myths and like); but far more strongly many of these writers suspect that the technology may destroy the content [1]:

> We see Ultratech is a very global idea because there is no language, there is no plot, there are no characters.... It's all just sound and light, TV taken down to its essence.

With this apprehension, we see a cousin—that artistic sensibilities and talent are replaced by technological prowess [2]:

> With the advent of electronic media, the material is abolished to all intents and purposes. Technique is no longer the ability to manipulate material, but the ability to manipulate technology.

Many of these writers and artists are sensitive to the Siren of the New Toy: the flashing lights, the stunning sound, the fabulous images.

But we should remember the play on technique and technology that art has always enjoyed. Haydn showed us sound, pulling musical jokes out of a serious score; painters have pushed color, shape, and the very technology of canvas into our faces, just as playwrights from Shakespeare and before have contrived to announce their medium as such. Media have always enjoyed self-reference, the sideways flip from surface decoration to the harmonics of meaning. It's an activity that necessarily shoves the media, and technology of performance, up in front of the content—at least for a bit.

Are the new technologies of pure digital sound, virtual reality, enormous screens, and computerized performances so vastly more overpowering that artistic talent will vaporize, like a candle gleam before the exploding light of a nuclear device? Will the surface scan of these computer screens blast away the references to life, to the mundane, that "art" has usually encompassed?

Of course, radical changes in technology provide new metaphors, new ways of thinking about the world, humans, nature, and the links between these.[2] And so new art is inevitably generated by these changes.

But some of the contributors to this volume suspect that computer technology may not be taking us to new art, but to no art. They note that computer technology powerfully tends towards the self-referential, and

---

2. See, e.g., [3].

(in a rapture, the ecstasy of narcissism) that technology obliterates the real world. In this way, technology dispenses with "meaning" in the sense of making or illuminating a link to "truth": to the world, humans, life, and so on. In that digital Hell, the body is turned "into nothing more than another station in the mesh of communicative networks" [4].

Perhaps only for the susceptible. We are, for now, fascinated—but only for now. People likely looked at the output of the printing press in Gutenberg's time as technology before they read the manuscript for content; folks remarked the fact of the new "talkies" sound films before they simply heard the movie as a complete audio-visual work. It was true for the sound of the first telephone voice. As I've suggested elsewhere [5], the graphite mark of the first pencil probably fascinated its first users, who exulted in its soft thick penumbral medium.

The articles in this collection reveal us as early initiates, those to whom the power of digitization is first given. They have an infectious enthusiasm for, and a lingering distrust of, the new works of technology. In retrospect a decade from now (and nostalgia aside), I hope we will not be too distracted by the then-ancient technology.

References

[1] Hayward, Philip, "Industrial Light and Magic," quoting Jonathan Klien, "Cool Art of Wild MTV Idents," *Invasion*, March 1988.
[2] Welsh, Jeremy, "Power, Access and Ingenuity."
[3] Bolter, J. David, *Turing's Man,* University of North Carolina Press, 1984.
[4] Murphie, Andrew, "Negotiating Presence."
[5] Karnow, Curtis, "The Electronic Word: Democracy, Technology, and the Arts," Review, 16 *Hastings Comm/Ent L.J.* 501 (1994). (In this volume as "Transfixed by the Electron Beam.")

15 Implementing the First Amendment in Cyberspace

As the inheritors of the Constitution, the honorary descendants of those who rebelled from the English Crown, we believe that we have rights—civil rights—that guard us from the arbitrary exercise of power and against the evisceration of individual prerogatives to privacy and unfettered speech. The denizens of cyberspace would have those rights—real rights, not imitation or virtual rights—govern actions in the electronic fora as well as in the more mundane and familiar contexts.

But this debate is infected by strange assumptions: some write off the electronic medium as obviously beyond the purview of 18th-century guarantees and cannot conceive how (for example) the right to be free of unreasonable searches and seizures could possibly apply to the electronic medium; on the other side, some assume all rights enjoyed by bodies in the conventional world necessarily pertain to our electronic interactions. The debate is aimlessly fueled by lawyers uncertain of technology and console cowboys who have never read a Supreme Court case interpreting the Constitution.

With a slim toehold in both camps, I will outline below some of the issues involved. Limiting myself to First Amendment issues, I briefly trace the development of some aspects of First Amendment law, look at its current applications in the networked world, and finally extrapolate to the broadband interactive immersive communications environment we call virtual reality.

EARLY DEVELOPMENTS: REGULATING THE ACTIONS OF THE GOVERNMENT

Broadly speaking, civil rights come in two general flavors: (1) the procedural rights to fairness, including the right to equal treatment, and (2) the substantive rights to free speech and to freedom from certain kinds of searches, seizures, and punishments. These rights are not reflections of divine prescription, even if such ruminations informed the thinking of the country's founders; these "freedoms" are generally just restrictions on the activities of government.

At first, following the adoption of the Bill of Rights in 1791, the proscriptions (such as those against unreasonable searches and seizures found in the Fourth Amendment) related to action by the federal government only;[1] state governments were bound by state constitutions, but not by these federal bars. The Civil War laid bare the fatal error of relying wholly on states fairly to govern the rights of their citizenry; in 1869 the Fourteenth Amendment followed, a direct constitutional proscription on *state* action. The Fourteenth Amendment's guarantee of "due process" led in fits and starts to the so-called incorporation of much of the Bill of Rights (the First through Tenth Amendments),[2] thus enforcing these prescriptions against the states, as well as the federal government. By the 1930s, a series of decisions from the U.S. Supreme Court had endorsed, as against *state* agencies, the First Amendment's guarantees of freedom of religion and expression.[3]

But all through this long process, the principle remained that the Bill of Rights acts as an inhibition on *governmental* abuse; the First Amendment does not directly govern relationships between private individuals. Taken in isolation, the First Amendment is not implicated by a person's rude, obscene, or otherwise offensive language, or a person's insistence that certain things not be spoken (such as a parent's request that politics or alternative lifestyles not be discussed with aging Auntie Cathy), or indeed an insistence that a specific religious formality be made. These are private actions, without governmental coercion or the imprimatur of a public agency.

The looming presence of the state is, of course, of central import when laws are passed which *do* impose these restrictions. When state law allows a person to recover millions of dollars in a libel suit—laws by which a state court plainly punishes certain types of speech—state action is implicated; when the state regulates discussion of politics or race, or

1. "Congress shall make no law..." the First Amendment begins, plainly referring to the *federal* congress.

2. In 1947 the Supreme Court almost, but not quite, concluded that the Fourteenth Amendment required the Bill of Rights to be applied to the states—the Court was one vote shy on the issue, *Adamson v. California,* 332 U.S. 46 (1947). The first eight amendments were then "selectively" incorporated as against the states over the next 20 years or so [1].

3. See, i.e., *Near v. Minnesota,* 283 U.S. 679 (1931); *DeJonge v. Oregon,* 299 U.S. 353 (1937). See generally [2].

when it insists on prayer in the school, or prohibits (or interferes with, or even burdens) indecent speech—then, and only then, may the First Amendment weigh in. State action, as the phrase goes, is made out.

PRESENT CIRCUMSTANCE: THE FUSION OF PUBLIC AND PRIVATE

Does the First Amendment have a role in the electronic network? Is the network a public forum, a governmental entity, a place where restrictions as exist are imposed by "state action"? When an online service such as IBM-Sears' Prodigy bars certain speech because it might be inflammatory, do we judge the action by First Amendment standards—which might rule the ban illegal—or by the rules of etiquette and social custom that govern our *private* relations?

One short answer—the least complicated analysis—is that online services are by definition private enterprises and can impose any rules they wish. Users have a choice of systems, and can opt in or out of the contract offered by Prodigy, for example. In this view, 10,000 people communicating though a Net is qualitatively no different from two people on a phone line.

The analogy does not quite fit, of course; AT&T (or Sprint) does not interrupt one's phone conversation and place restrictions on content, and by the same token one's two-party phone call is not as—let's say it—*public* as the exchanges on systems such as Prodigy, CompuServe, and others.

But there are deeper problems here. One nameless acquaintance of mine informs me that the "privatization" of certain Department of Energy (DOE) nuclear recycling concerns is a sham: the companies are being sold, true, but the DOE is the chief stockholder in the new companies. Were the state to contract out its police function, no one believes that the private security force would be immune from Constitutional scrutiny. And difficult issues are posed when evidence that would be deemed illegally seized by a police officer is seized by a private citizen and immediately turned over to the police: it all depends on whether the police knew in advance or perhaps suggested the seizure. There are some very thin lines here.

In the First Amendment area, the lines wane so thin they can disappear. In a series of cases in the 1960s, the Supreme Court found that certain technically private areas such as shopping malls were *functionally* similar to public fora. The Court ruled that demonstrators had to be allowed to express their positions on the "private" property.[4] Protesters

4. *Food Employees Union Local 590 v. Logan Valley Plaza, Inc.*, 391 U.S. 308 (1968).

thus enjoyed the same rights to pass out leaflets to shoppers as, for example, to travelers at city-run airports. With a change in personnel the Court severely cut back on this doctrine,[5] and recent cases cast further doubt on the role of the "public forum" doctrine.[6] But the principle, in broad strokes, is probably intact: it is most unlikely that a company-owned town could stop the free expression that must be allowed by a municipality.[7]

In the Net as it is presently constituted, the lines are even more difficult to draw than this. The National Science Foundation runs a big chunk of the Internet, funds many of the Internet's nodes, and initiated the Internet's backbone NSFNET. Other federally funded and operated military, space science, and energy agencies run parts of the Net, many of which in turn rely on private companies (IBM, MCI Communications, Sprint, and so on) to do much of the work.[8] Government agencies really do set policy on large portions of the Internet, barring, for example, certain types of commercial "speech" such as advertising and certain commercial solicitations.[9]

As the last series of corporate names suggests, the picture is even more complicated. Privately owned telecommunications companies, inextricably involved in the Net, operate as quasipublic agencies themselves. That status can carry with it government-type obligations to require communications access, bar content regulation, and, at times, impose an obligation to allow competing positions on issues to be heard.

---

5. *Hudgens v. NLRB,* 424 U.S. 507 (1976).

6. With respect to property that was indisputably public—an airport—one faction of the Supreme Court ruled that a religious group's repetitive solicitations of travelers could be barred, a restriction that did *not* need to be tested against the standards of the First Amendment. This followed because airports were not meant for public discourse and did not have a "time honored" tradition of being so used, *International Society for Krishna Consciousness, Inc. v. Lee,* No. 91-155 (June 26, 1992). Another faction of the Court held, on the same day, that such reasoning was in error: that governments could not be allowed to decide what fora were, and were not, meant for public discourse and found that handing out leaflets in the same airport was *protected* under the First Amendment, *Lee v. International Society for Krishna Consciousness, Inc.,* No. 91-339 (June 26, 1992). The serious division in the Court reflects profound confusion as to (1) the meaning of the "public forum" doctrine, and (2) how to treat the property interests juxtaposed to the rights to free expression. It may be enough to note here that the Net, and its expected descendants, are plainly meant for public discourse.

7. *Marsh v. Alabama,* 326 U.S. 501 (1946) (state law cannot allow "company town" to bar distribution of pamphlets).

8. See [3,4].

9. This is the NSF's "acceptable use policy," the subject of testimony—and opposition—by Mitch Kapor before the House Committee on Science, Space and Technology. Testimony re: NSFNET and future of the NREN (3/12/92). See generally [5].

The imposition of that sort of regulation generally stems from one or two principles: (1) operating a monopoly carries with it the responsibility to operate it in the public good, which cannot be otherwise assured by the usual response to competition, and (2) utilities providing public essentials (such as power and water) must be regulated to ensure the essentials are provided.

In the electronic world, however, the courts' approach has been based on ancient facts. While the factors outlined here might suggest that there is sufficient "state action" on which to insist on true First Amendment standards in the electronic media, the courts have deferred to the regulations made by agencies, such as the NSF and the FCC, on the basis that those agencies know best how to divide up the meager resource called bandwidth.[10]

The consequence can be severe regulation, content-based discrimination, and outright bars to certain types of speech, all in the name in fairness, ensuring access, and the husbanding of a scarce commodity. But the "commodity" is actually expanding exponentially: the number of cable channels is growing to the many thousands and interconnected PCs are in the millions, all as the relative importance of print media—in which content-based regulation would be anathema—declines.

THE RATIONALE OF RESTRICTION

The forces arrayed in favor of continued and, indeed, stronger regulation of the electronic medium cannot be underestimated; virtually every reader of this article shares some of those interests. Agencies and legislatures place serious restrictions on speech, sometimes for good reason. A host of laws punish certain kinds of speech, regardless of the medium, such as threats to kill the President of the United States, false advertising, slander, and libel. Other laws allow consumers to control the despatch of personal information; they allow people to sue for breach of privacy, for stolen trade secrets, for uttering privately owned speech (such a copyrighted poem or play), or for sounding too much *like* someone else.[11]

---

10. See, e.g., *Columbia Broadcasting System, Inc. v. Democratic National Committee,* 412 U.S. 94 (1973); *F.C.C. v. League of Women Voters,* 468 U.S. 376 (1984).

11. Singer Tom Waits successfully sued when someone else made an advertisement with his distinctive low, raspy, growly, gravely, strung-out voice. *Waits v. Frito-Lay, Inc.,* 23 U.S.P.Q.2d 1721 (9th Cir. August 5, 1992).

All of these are plain inhibitions on speech, created by governments, but most of us would endorse most of the regulations most of the time.[12] But three areas seem to stand out as especially vexatious in the electronic-telecommunications medium: privacy, the related area of security, and moral issues. All offer significant arguments for regulation, or to use stronger language, censorship.

First, privacy. Many of those arguing for the augmenting of civil rights, and against censorship, are also in the vanguard of the privacy rights movement. These persons are, with very good reason, especially concerned that the agglutination of Nets combined with the escalating digitalization of information poses serious risks to privacy. Motor vehicle registration information, debt, income and asset information, together with buying patterns, associations, travel history, and political contribution history, at least, are all available somewhere in cyberspace. Laws and regulations occasionally bar use of some of this information; many argue for stronger measures. When the Lotus company thought of combining some of the available consumer information on a CD-ROM for public distribution, vociferous and widespread opposition broke out, and Lotus set the plan aside. However, the information is there, still, for most anyone who wants it.

Second, security. At this point, concerns with respect to security are still traditional: to protect privacy and to protect against data loss, either by theft or digital viral infections. Increasingly, however, a third problem is of significance: attacks on computers running real-time systems.[13] This goes beyond the loss of data, which, theoretically, can ultimately be replaced; interfering with a subway system, an aircraft in the air, or certain manufacturing systems results in physical damage and loss of life. Real-time injury becomes increasingly likely as the Internet grows, increasing its interconnections, and maturing those connections far beyond electronic mail. This risk will cause systems operators and communications providers to block access and inhibit information exchange.

Third, morals. The informal censorship known as the "politically correct" movement, based mainly on some college campuses and occasionally mocked in weekly mainstream publications, is nevertheless part of a larger issue. Violence, patent sexual innuendo (i.e., something less

12. There are scores of difficult, interesting grey areas here, where First Amendment rights conflict with other rights; but generally these are not peculiar to the electronic medium. For example, trademarked materials may be used by others without permission under certain circumstances, *New Kids on The Block v. News Publishing, Inc.*, 745 F.Supp. 1540 (C.D.Cal. 1990), aff'd., 971 F.2d 302 (9th Cir. 1992); "threats" to public officials sometimes are more strongly spoken political invective and less true threats to mayhem; and so on.

13. See generally [6].

than obscenity), vicious attacks on various lifestyles, blatant racism, and anti-Semitism—these are evils that various significant portions of the citizenry would bar. In regulated environments, substantial pressure is placed on agencies and individual broadcasters to stay away from such odious content, and the pressure comes from all areas of the political spectrum. Where a Dan Quayle objects to unwed mothers, a liberal parent might as forcefully object to constant murders depicted on the television. Reasonable people differ on the line between illegal hate crimes on the one hand and political and social invective on the other, which the First Amendment requires that we tolerate.[14]

We are increasingly aware of the thinning line that divides the environment from self; the news and entertainment media play an increasingly significant role in our daily lives. As media go interactive, and as the machines of high bandwidth spread throughout our homes, it is reasonable to assume that the rules on which we insist in our daily lives—the moral rules by which we judge and think of ourselves—may be mandated in our virtual lives as well. At least, there will be strong temptation to make it so. And every rule of opprobrium chills the right to expression.

FORWARD: TRANSITIONS FOR THE FIRST AMENDMENT

High-quality kinesthetic experience is likely to result from current developments in three-dimensional sound, advances in digital signal processing (DSP) chips, central processing units (CPUs), and cheaper and higher capacity random-access memory (RAM) and mass storage devices, all anticipating increasingly realistic immersive video. The narrowing of the difference between reality and virtual reality will make it the more imperative that the constitutional rules governing our lives in the outside world be applied to our electronic doubles.

At the same time, we have the continued involvement of government in the creation of the high-speed national and international electronic backbone to be known as NREN. Bandwidth providers regulated by both state and federal agencies will increase their involvement beyond telecommunications, enlarging their interest to the direct provision of data and interactive video to the home.[15] Net-based communications and

---

14. In *R.A.V. v. City of St. Paul,* 112 S.Ct. 2538 (1992), the Supreme Court found a state statute which banned so-called "hate" crimes in violation of the First Amendment.

15. Readers may recall AT&T's financial stakes in the upstart home video game company, 3DO [7], in McCaw (the biggest U.S. cellular operator), EO (mobile, handheld computers), and General Magic (software for interactive mobile devices) [8].

interactions will become more widely available and increasingly used as a public arena. High-speed interactive immersive communications—virtual reality—will be an inseparable function of private and public assets, just as the Internet now is an indivisible, constantly mutating, melding of private and governmental machines, networks, cables, satellites, software, and data.

In short, the line between public and private will be eviscerated. This leads to two juxtaposed results: a license for more regulation and a stronger basis to object to the regulation on First Amendment grounds.

As the technology becomes more opaque to judges only schooled in the law, the courts may increasingly defer to the rulings of regulatory agencies. Extrapolating from the history of staunch regulation in the telecommunications, radio, and television industries, courts are likely to distinguish away from the "no holds barred" First Amendment approach used with the print media. Regulation will be tempting: if legitimate interests in privacy, security, and morals cannot be enforced in cyberspace, the argument will run, they cannot meaningfully be enforced at all.

But the burgeoning Internet is not the medium of a few licensed broadcasters, but rather an interactive world where the entire citizenry can meet, exchange information, debate issues, and make objects of art, all in a milieu shot through with governmental participation. In this context, the deeper forces animating the First Amendment are likely to prevail.

Real-world restrictions on expression are legitimized by the need to protect property, broadly construed. Reputation is property, and so a slander suit punishes some speech; trade secrets, copyrights, and trademarks are property,[16] and so theft and infringement—certain types of expression—are punished by law. In the public forum area, it is often the law of trespass—designed to protect property—that is used to block free speech.

These property interests, including the privacy interests identified earlier, can be thought of as expressions of a distinctive right: the right to do as one pleases with what one owns. The right to free speech finds its limit in the vicinity of another's right to do as he pleases with his property.

In cyberspace, these countervailing property rights are of uncertain dimension. A full discussion of their transmutation is beyond this chapter, but it is reasonable to predict mass authorship of objects and, as the result of constant mutation by anonymous authors, the gradual breakdown of the concept of recognizable objects capable of ownership. Places, utterances, objects, and personalities will be as much a function of the ground or environment (which belongs to everyone) as a single individ-

16. See generally [9].

ual's efforts.[17] To picture the context, the slander of an electronic persona—a large red lobster, say[18]—does not appear seriously to impinge the rights held by people in the real world.[19]

In short, there will be a far weaker rationale for speech restrictions in cyberspace than in the real world. Unfettered by these restrictions, it should become increasingly plain that our electronic interactions are privileged—as privileged as speech on the street, because the Internet is a public thoroughfare; as privileged as the printing press, because on the Internet we are all broadcasters, reporters, and the public.

To be carefully distinguished, however, are acts in the electronic medium which are the functional equivalent of acts in other media (i.e., where cyberspace is simply being used transparently to modify the real world). Regardless of the medium, we will always punish slander, death threats against a real person, invading a real person's privacy, or tampering with a real air traffic control center's computer.

For those who delight in subtle distinctions, the future will be an interesting time. Difficulties between public and private in First Amendment law will fade as the concepts fuse, to be replaced by the analysis of the strange, blurred edge between the virtual and real.

References

[1] Tribe, L., *American Constitutional Law*, 2d ed., Minneapolis: West Publishers, 1988, p. 772.

[2] Cox, A., *The Court and the Constitution*, Boston: Houghton Mifflin, 1987.

[3] LaQuey, T., *The Internet Companion*, Reading, MA: Addison-Wesley, 1993.

[4] Lo, Catherine, "Get Wireless," *Wired* 142–147 (April 1997). (A list of "private" owners of the "public" electromagnetic spectrum.)

[5] Krol, E., *The Whole Internet User's Guide and Catalog*, Sebastopol, CA: O'Reilly & Associates, 1992, p. 31.

[6] 4 *Info Security News* 33 (March-April, 1993).

[7] "Virtual Reality Gets Real," *The Economist* 61 (Feb. 20, 1993).

[8] "The Computer Industry," *The Economist* 18 (Feb. 27, 1993).

---

17. Tom Barrett of Electronic Data Systems may have been referring to this development in his discussion of evolving computer interfaces. Immersion in what Barrett calls "infoplaces" will create the *environment* as the message (content). That is, the environment will in large part be the information processing facility (DPMA conference, reported in [10]).

18. If I recall rightly, a "life-sized" [sic] lobster was used as a virtual-reality incarnation by the operator in an early Autodesk demonstration.

19. As the text suggests, there are perplexing consequences for traditional concepts of copyright and other types of intellectual property, which have always assumed fixed, identifiable objects owned by one person or by a limited, identifiable group of real persons. That is the subject of Chapter 5, "Data Morphing: Ownership, Copyright, and Creation." of this volume.

[9] Nimmer, M., and D. Nimmer, *Nimmer on Copyright,* § 1:10 [A], 1992.
[10] 3 *CyberEdge Journal* 4, 6 (Jan./Feb. 1993).

Transfixed by the Electron Beam 16

[G]eneral theories, made to fit everything, tend to fit nothing in the end....Universal theories are prestigious, and they give an illusion of power. In fact, very often they're weak and ineffective.

—Alexander Tzonis [1]

How wonderful it would be if the computer revolution sweeping the world would at once solve the troubling problems of academia; resurrect the humanities; allow radiant creativity to flourish as never before and in the hands of people with absolutely no talent; and at the same time prove that one's own specialty, long despised by one's peers, was truly and really the central, most profoundly relevant discipline of them all—able at once to comprehend the others and more than any other pursuit integrate future technology with the West's Greco-Roman, Judeo-Christian, Aristotelian heritage.

Well, *Lanham has done it.* He is a professor of English (rhetoric) at the University of California at Los Angeles, and he has discovered *computer fonts* [2].[1] This discovery (and all it entails) compares with Lourdes, and the transformations it promises are as miraculous. Professor Lanham is convinced that the great classics of the West are comprehensible only by those schooled in rhetoric and that the computer brings rhetorical de-

This chapter is a review of Richard A. Lanham's *The Electronic Word: Democracy, Technology, and the Arts* (Univesity of Chicago, 1993).

1. Parenthetical number references in the balance of this chapter, such as (5), refer to pages in the book reviewed here.

vices squarely to the forefront, able at a single blow to provide the basic analysis of legal thinking (64), literature, art in general, management theory (66), game theory (69), teaching and the university, chaos and nonlinear systems, war and peace (83), and on and on. Rhetoric "draws all subjects into its political and social structure" (112, see also 63).

This is one *heck* of a college major.

What is rhetoric? A perusal of the 10 overlapping articles in this collection reveals that rhetoric "resembles what is now called in many fields, a nonlinear system" (61), now "at the center all across the disciplinary landscape" (63), "the art of persuasion" or the "economics of human attention-structures" (227). By his own repeated admission, the field has been "discredited" (114), but Lanham now has his revenge, for the onslaught of digital data, widespread computers, and the ease with which we may now all create and mutate extant texts all reveal the truth of rhetoric's claims and place it back on the throne: it is now a "general theory for all the arts" (16).

This review, of course, is rhetorical, as is everything else. So Lanham defines himself to be tautologically correct. All communication is subject the rhetorical label, and most human actions have some communicative feature—a convenient equation. How better to bootstrap one's field than strap it onto the tide of technology, claim the technology for one's own, and so share in the rising flood?

What, then, is Lanham's loop between computers and rhetoric? What is the stunning link he has discovered between rusty ancients and the gleaming chrome mirrorshade future? This: computerized digital text is now "malleable and self-conscious" (5): one can look first "AT it and then THROUGH it" (5).[1] Is this clear? Before, text was passive: predigital text (print) "wants the gaze to remain THROUGH and unselfconscious all the time" (43). Now, because one can both read the text *and* then manipulate it as an object, text is now permanently "bi-stable." (Look at the stunning font changes! Look at the edits one can make and *still have the text appear as an original.*) Looking THROUGH the text (presumably to its meaning and direct comprehension of its referent) is "Newtonian" and "Arnoldian"; but look AT it, in the way computers allow, and we have "deconstructed" the Newtonian world into the rhetorical one: a view of the communication itself as the object of study.

But as Lanham's own examples make quite plain there is nothing new about this "bi-stability." The arts of persuasion, as Lanham wants to make perfectly clear, date back to the Socratic dialogue and before, and while he doesn't belabor the point, the arts ever since have paid homage

1. The capitals are Lanham's and he uses these, in almost all the essays, every time he prances out this piercing analysis.

to it. And, I suggest, so have the rest of us, back through the Dark Ages.[2] Every illuminated manuscript, every poem, and every prose sentence with pretension to anything more than the dullest exposition of information aspires to an elegance done with the exploitation of form, rhythm, meter, and rhyme. The form/content dualism is old, old stuff, and Lanham is not well served by dishing it up again with a nifty little phrase.[3]

But in fairness, Lanham does suggest that there is something, *something* new here, sufficient to reinvest rhetoric with a new *justificandum*.[4] The "AT/THROUGH oscillation," while it is clearly a postmodern phenomenon (46) and also a *pre-postmodern* phenomenon, maybe a 16th-century phenomenon (159), and an ancient phenomenon, too (51) (after all, Aristotle is no spring chicken), has been given an extra boost, made more vivid as it were, by the central fact that "the arts, through their common digitalization, have become directly interchangeable" (77). In digital media there is "no final cut" (7), and all media is thus "*user definable*" (15; Lanham's emphasis). This "user" power is exerted through the zoom box (108)—making the "big decision" of what size the text box is to be (41) because to "change scale is to transform reality utterly" (42). In this way, digital manipulation makes the bi-stable oscillation *more obvious:* computers destroy the artifice, the illusion, the pretense (under which we have been living all these centuries) that there is no bi-stable oscillation (81-82)! Now we see—and can never again evade—the truth that language is *both* ornamental and purposive. "This is a toggle to boggle the mind" Lanham mischievously interjects (82) and then observes that, while we have all along been hiding from ourselves and from this fundamental oscillation, the electronic revolution—electronic text, as such—reveals the "fundamental renegotiation now demanded of our Western reality" (82).

2. The Gutenberg Bible itself was—and is—a thing of great beauty, made with as much attention to presentation and ornamentation as to the faithful reproduction of the content—a faithfulness accomplished by reproducing in movable Gothic type the calligraphy of Johann Gutenberg's time. The type, now for the modern reader, is almost impenetrable; there is enough ornament to cheer any rhetoritician. By 1475, the Mainz Psalter had been published, "studded with vividly detailed initials" [3].

3. A wonderful look at fonts, meta fonts, and the old problems of representation is found in [4]. See also [5], which discusses the decoration of both computer and precomputer "fonts."

4. Lanham seems to prefer words like *explanandum* when he might have said explanation or something along those lines (17). Perhaps using Latin is a rhetorical device, calling attention to one's techniques; perhaps it is homage to the classic texts, although (as we see later) the balance of Lanham's writings put that intent in doubt. Speaking for myself, *Ecce, denuo ago, sicut soleo! Non enim possum facere quin Latine loquar* ("Oh, darn, there I go again! You know, I can't help speaking Latin"). Courtesy of [6].

The scales are lifted; our security blanket is gone. The Great Lie of Western civilization[5] is made right. The mouse and zoom box will change reality—they reveal the truth of our ontological schizophrenia. Really (81, see also 63, 75). This is the big time: not even the lawyers involved in the *Apple v. Microsoft* litigation[6] thought so much was at stake.

Is it so? Is it true that the computer and "electronic text" over the last "twenty years" have transformed and resolved the form/content or surface/depth dualism in writing, the arts, and teaching (63, 85)? In the following sections, I look at Lanham's two major thrusts: his discussions of (a) the new changeable surfaces presented by the computer, and (b) the new impossibility of teaching authoritative texts as the result of digital information.

FALLING IN LOVE WITH THE MACHINE

"Perhaps," Lanham muses, "the most widely debated, though far from the most important, issue involving electronic text is whether writing on a computer creates verbal flatulence or not" (40). I will admit I hadn't know the problem was so widely discussed, but after hacking my way through Lanham's essays, I see it all in a new light. His computer should be put to the flames: he is too enamored of the thing, fascinated by the fillip of fonts, too merry with his floppy drive.

No doubt digital media promises changes, and towards the end of this chapter I refer to two of the more interesting shifts Lanham mentions. But Lanham proves that remarkable exaggeration is still possible even in this hype-laden environment. It stems, I think, from his fascination with the process of computing, the fingertip manipulation available to everyone who can afford a machine: this harbinger of a new kind of worldwide democracy. He forgets that in the grand scheme of things few have access to computers, and of those that do, the vast majority use them to get a job done: pump out a paper, create the annual report, write checks, design an engine—all tasks that were done, albeit slower and more clumsily, before the PC arrived just over a decade ago.[7] For these people—we call them the

5. I am *not* making this up. Lanham refers in this context to the "carefully tuned illusions from which Western social reality has always been constructed" (43).

6. The litigation involved, among other things, a copyright dispute over aspects of the Windows graphical user interface. See *Apple Computer, Inc. v. Microsoft Corp.*, 799 F.Supp. 1006 (N.D. Cal. 1992). Near the end of his last essay, Lanham refers to this litigation, citing to the "concept and feel" (sic: look and feel?) of issues litigated there (276). See generally [7,8].

7. Telecommunications and its associated network technologies stand out as an exception here. Computer-mediated communications, from email to MUDs—where people meet and role play online—to the interactions of the Internet and the synthetic worlds of the

consumers, the economic rational and life of the computer "revolution"—there is no joy in the nice semiotic distinction between a macro, a program, text, data, and the algorithmic equivalencies of digitized music, text, and art. Many of them don't even like the machine.[8] But for Lanham, the computer and its screen loom hard on his horizon; he's lost in the machinery—like a car mechanic who revels in the grease, the beauty of the gears, the smooth luxuriant curve of steel and chrome.

His is the fascination for the new toy. Before the new technology and all the new terms (chaos, complexity, hypertext, critical theory, interactive fiction), Lanham is a deer transfixed by the bright electron light of an oncoming machine. It is an interest in an immature technology still finding its place in the humdrum world, a romance that calls to mind those daring young men in their driving machines of the early twenties, fixated on getting the damn things to work. The mass of humanity today, on the other hand, uses cars to get someplace.

Computers, like cars, are likely to fade to background. While we are indeed still forced to confront the computer as tool and thing-to-be-figured-out, the machines will "eventually get out of the way of the work" allowing the user to just do what needs doing, according to Mark Weiser of Xerox Palo Alto Research Center (PARC), who ought to know.[9]

To the extent that computers do not do this—that is, to the extent that the interface fails to be a transparent medium—serious problems can arise: the user can be inhibited, error creeps in, and "the operator's perception and experience are *of* the computer" instead of the work that is to be done [15]. Professor Sheridan's work makes the point that what Lanham calls the degree of "AT/THROUGH oscillation" depends profoundly on the context, the user's needs, intentions, facilities, habits, familiarity with the system, and the role of other humans in the system. This is not an odd conclusion: what is "natural" for one is not for another, just as a

---

future all promise a qualitative shift. See generally [9]. Telecommunications and networks are not the topic of Lanham's essays, except once when speaking of public electronic personas. I note this at the end of this review.

8. Technophobia afflicts about 55% of the U.S. population. Twenty-five percent have never used a PC, programmed a VCR, or set up a favorite-station tuning on a car radio [10]. See also [11] (fear as response to new technologies).

9. PARC brought the idea of windows and handheld mice (which Lanham so admires) to the computing world and remains one of the leading computer research facilities in the world [12]. Computers disappear when, for example, they dissolve into washing machines and other appliances, when they are integrated into touch-sensitive bulletin boards, and voice recognition and reproduction achieve high quality. See [13]. See also the interview on virtual reality with Dr. Thomas Furness [14] ("I've learned how powerful a medium this is. This immersive environment, this 'circumambience' of visual, auditory, and tactile information, gives us the opportunity to—in essence—do away with the medium...the medium disappears").

font easy for Gutenberg to read is gibberish to most of us.[10] This undermines Lanham's generalizations on the "power" of the "new" graphical user interfaces with their zoom boxes and other tools. Whether digital manipulation of text reveals the dualistic nature of the word depends; it depends. It might for a while and for some; and then later not at all. No one now concentrates much on the graphite of the pencil, although I expect it was a wonder once upon a time and could be again in the right hands.

All "new" technology draws attention to itself; perhaps it is novelty, and not computers, that informs Lanham's view of the screen as window and wall.

A central problem with Lanham's analysis is that he doesn't know very much about computers or of such related fields as complexity theory, chaos, and the like. He may well be the one-eyed king in the land of his blind "Luddite" fellow humanists (23), but his information is very much second-hand stuff. He insists, for example, that the "complexity movement...has a lot in common with the rhetorical way of thinking" (251) in an essay that is prefaced by a concession that he cannot follow information theory, chaos theory, or complex systems (225, see also 151). He remains convinced, however, of "astounding correlations" (226) because all of these new theories are coevolutionary, bottom up instead of top down, and resonate in harmony with early pre-Socratic Greek sciences and (oh, yes) "Eastern religions as well" (226). When Lanham tells us that hypertext is "chaotic" and "totally nonlinear" because a reader's path through the text cannot be predicted (94, see also 71), he does himself no service at all; with that remarkable coincidence, everything is in common.[11] Lanham has read a bestseller on chaos [18]; good, but we cannot trust Lanham's judgment here. And inevitably the suspicion roils like smoke over the whole battleground: We're not sure what he knows about the other stuff, either.

---

10. See above, notes 3 and 4.

11. Chaotic systems are highly sensitive to initial conditions. For example, slight differences in billiard shots produce radically different outcomes. See generally [16]. Systems with very few elements, such as three interdependent pendulums, and simple uniform systems, such as gases, can exhibit such chaotic behavior [17]: Over time one notes an increasing number of system configurations stemming from arbitrarily small differences among initial states. "Chaos" applies to dynamic systems, detailing the behavior over time of interacting components. The issue is not, as Lanham seems to suggest, a matter of many pieces (as language has many parts)—a "complicated" puzzle with 10,000 interlocking parts exhibits no chaotic behavior—neither does the availability of multiple paths through time in hypertext (71, 94). An English shrubbery maze is not "chaotic."

DESTROYING THE UNIVERSITY IN ORDER TO SAVE IT

The second central theme of the chapters in this book is the survival of the humanities in the university.

Lanham suggest that he has found the synthesis, or perhaps a middle ground, in the *Great Books v. Relevant Books* debate: "that fiery argument is pointless" (150). Under the influence of mass immigration, new languages, and the introduction of traditions other than those of the standard Judeo-Christian culture, "we are required to find really new ways to widen access to the liberal arts without trivializing them" (103). Because electronic text allows a bi-stable self, *both* world views—the classics on the one hand,[12] and the "interactive" taught with "understanding and zest" (106) on the other—can be preserved (260).

There is a hint of promise here: the struggle he notes is, I think, real: teaching is forever a task of pulling back the timeless texts into the present time. This is very, very hard, and so there is a profound difference between good and bad teachers. How *does* Lanham propose to widen access to what he calls the "non-traditional student" (106) and yet not trivialize the liberal arts?

The answer is this: he proposes to trivialize them. The interchangeable digitalization of data allows anyone to do anything—it's "now almost a do it yourself affair" (106). This what "democracy" means in the subtitle of this book. Anyone can make music (106), a painting (107), a terrific new *happy* ending to King Lear (6)![13] Lanham does not quite say that no talent is required—he says that all this can now be done by "people who before had not the talent" (107). Do not ask what is "good." Any of this music can sound "good" Lanham says, invoking the hoary ghost of John Cage and his cacophonous prepared pianos (12). Here's what Lanham praises: John Cage's rapping a closed piano with his knuckles—once or twice—which "Nam June Paik dramatized more vividly by taking an ax to the whole instrument" (17). Let us have "no invidious discrimination between high and low culture" (14). Goodness, no; I just hope we have enough pianos for everyone.

---

12. Both Lanham in his essay, and I in this review, take liberties with what we mean by the "Great Books." As Lanham notes, the academic debate as so phrased is fairly recent and, strictly speaking, refers to a certain select series of works (not Confucius, for example). But here, I use the term more carelessly, to cover the teaching of a canon of work more broadly construed to encompass works selected by authoritative teachers with the specific intent (perhaps among other intents) to present the works as manifestations of a culture.

13. As Professor Alfred Harbage (the editor of the Pelican Shakespeare) has noted, it was Shakespeare's particular genius (for the first time) to insert a tragic ending to the legend of Lear [19]. Never mind; happy is good too.

And in teaching, this is what Lanham proposes for the textbooks: through the magic of digital, infinitely malleable text in the hands of every student, we arrive at an "incredible personalization of learning, a radical democratization of 'textbooks' that allows every student to walk an individual path" (10). Teacher and student will collaborate on this (127). "Problem solved" (129). Well, yes. If every student can invent his own learning book, learning won't present much of a problem. With each person on her own track—everyone treated as "nontraditional" I suppose—the lecture course will soon be gone; so, pretty much, will others.

But note this. None of these students Lanham would so purport to teach would understand his book. Lanham, who loves the Greek and Latin quotation, the daring allusions to deconstruction and the cozy, familiar reference to Arnold, Aristotle, Plato, and hundreds of others in the liberal arts canon[14]—Lanham, where will he find readers out of the generations he "teaches"? *In forum, in Senatum, in conscionem populi, in omnem hominumum conventum*...(164)? Maybe, maybe. Perhaps this is nothing less than *he ton logon paideia* (155)? Hah. *Tertium non datur* (210). Lanham's students need not apply.

Of course, Lanham *likes* the Great Books (he's got that going for him), but the teaching of them is no longer workable (140): witness the multicultural student referred to earlier; witness the irrelevancies of the liberal arts to the "social purposes of our time" (117). Actually, there is a very wide swing here, between saying that the classics *cannot* be taught and that they can be repackaged into brightly colored interactive multimedia, which *could* be taught "without necessarily compromis[ing] the wisdom therein" (105). Lanham takes both positions through these essays.

Here is Lanham's path to uncompromising wisdom and the social purposes of our time: Education need not include instruction on how to write essays; forget spelling, we have spelling checkers and global search and replace features in word processors; invent punctuation, use emoticons (those cute, enormously expressive symbols like :-) (sideways smiley face) or don't use punctuation at all; but do insert colors into the text! Can't read? Never mind! Dispense with the ugly chastising notion of "remedial" programs and let the illiterate use sounds and images instead (129). "Upon this theory our new electronic nontext-textbook will be built" (127-28).

We end up with a sort of primeval ooze; a nonlinear text that has no beginning, middle, or end; no rules about verbal excess or "expressive

---

14. He's "addicted" to reading and doesn't mind the reader knowing his catholic taste. I don't mean too much of an *ad hominem* argument here, but you don't find Lanham educating himself much by morphing bit maps.

self-consciousness" (129). Just as good as Shakespeare: all one "single spectrum of expressivity, with no need to stigmatize any area of it" (129). What is left then of teaching? Of the university? Why attend? Any reason other than to pay the salaries of the professors?

In a nation with stunningly high illiteracy, where college graduates cannot read or write, much less read Plato with any comfort, where the arts struggle for survival and our cultural links become increasingly fragile, there is something truly horrible about Lanham's eager paean to fragmentation and ignorance.

Something in Lanham thinks that the university and its aspirations should be no better than its average environment. If they can't read when they get to college, then by gum they won't *in* college either; if students come from the four scattered directions and scores of backgrounds, then so shall they be delivered at the end of their term. And so believing that the world is too complicated, he surrenders (141):

> Our intellectual world is too various and volatile to accept an imposed conceptual coherence even if we could agree on one to impose—and of course we can't, one person's *Republic* being, as we have come to learn, another's *Thoughts of Chairman Mao.*

There are two things wrong with this instrument of surrender, quite aside from the ready and unbecoming ease with which it is offered. First, in the university, of all places, both Plato and Mao can be taught. And they are. Second, they can both be prescribed as authoritative texts that require mastery, not as pieces of data that can be incorporated into some nice database or glued into a multimedia production.

The difficulties of handling wide spectra of thought are not new to the academic: Mao and Plato may well have more in common than Edward Albee and Chaucer. The problems of dealing with a wide variety of in students is old, and did not erupt (as Lanham suggests) with any recent influx of immigrants.[15] Why is Lanham so ready to abandon ship? It is a lure of the computer, perhaps: it shows him that in its ability to reduce everything to the lowest common denominator (quite literally: 0111001110101110 110110) he can reinvent the field of rhetoric, the

15. Changing demographics were evident as the Pilgrims stepped off at Plymouth Rock. As Clark Kerr (previously president of the University of California) notes, the university has constantly dealt with the students who identify strongly with fractions of the "outside" community, who come from all classes and races of society, and are of all ages [20]. Certainly since the end of the Second World War, there has been an express recognition of the issue of maintaining the elite of merit which is the university, in an environment dedicated to an egalitarian ethos [21].

study of form, style, and techniques as the leading wave, and he can surf it to a new prestige.

But championing the lowest common dominator is a wretched thing to do. In Lanham's university, there would be no stigma, no judgment; one's creation, interpretation, analysis, and editorialization are one and the same, and equivalent to the action—any action, I suppose—of anyone else. The computer brings the "'democratization' of 'originality'" (46)(this is what Lanham *means* by democratization). Who is to judge what is "good"? No one (12).

We are all geniuses. Or none is.

This is the worst possible effort to manage the possibilities offered by computing. Rather than seeking out the means of augmenting the human imagination, leaping to new heights, Lanham urges a collapse to the dullest, most mechanical of outputs—whatever we can all do with mouse button clicks. The problem is that, caught up in the fad of the information revolution, Lanham sees everything through that lens: He is a man with a hammer, and all the world looks like a nail.

For Lanham, the new nontext textbook is an "open ended information system" (126)[16] distributed through fiber-optic integrated services digital network (ISDN) (yes, another pointless acronym).[17] But teaching is not exactly the transmission of information, except at the rawest level of analysis. We do not study the canon—whether the Great Books, Lesser Books, Freud, Mao Tse Tung, or The Buddja—to secure information. Shakespeare and Confucius do not "inform" us in the way numbers or graphs inform us; we do not read Plato or even a scientist such as Heinsenberg to evaluate the truth of what they say, to debate whether they were right or wrong. We do these things to generate a sense of context and history, to become articulate members of a society.[18] This is most obvious in the arts: the study of music or painting are not served at all by simply playing with sound generators and moving pixels about a screen with a mouse.

It is emphatically the role and place of the university to maintain the conditions for a shared culture, which is an historical entity.[19] The rheto-

---

16. "[I]n an information society...the words *are* the goods" (229).

17. Did Lanham just read about ISDN somewhere? Why not asynchronous transfer mode (ATM)? ATM is more expensive, much faster, and more fashionable now than ISDN. See generally [22]. Or how about satellite distribution?

18. But I repeat myself: There are only articulate members. Ignorance, and the inability to communicate, disenfranchises and destroys membership in the society.

19. See [23].

ricians have, for centuries, sought to devolve the university to a trade school, where more "practical" studies were taught [24].[20]

This is not enough—above all the task of the university is, in the face of endless fractionalizations and social diversity, to make a more unified and deeper intellectual world [27], to induce membership "in that city of letters not made of hands," the ancient company of scholars [28].

This transmission of culture is a constant remaking, but the remaking is done as one submits to the words, numbers, and text of the old makers [29].[21] "Authority untaught is the condition in which a culture commits suicide" [30].

The authority of the past work is a prerequisite to intelligent play with the medium. There are, to be sure, a variety of approaches to those works, but all take the works as authoritative. Those authorities are our common ground; once internalized, they allow us to communicate to each other. Without them, we will just make noise at each other and do beastly things.

This authority is exactly what Lanham denies. Earlier, I presented Lanham's suggestion that he mediated the Great Books debate, but he lied, and so did I. Lanham comes out squarely against the Canon; in the same spirit as his joyous abandon of essays, spelling, punctuation, and the rest, he urges the abandonment of training in print—rather he says, train our children for the electronic media (xi). Well, not my children. Yours? If Lanham were right, all key educational goals could be achieved with individual students locked up in rooms with computer terminals and a feed to the university database. No exam would test for, and no value would be placed on, a student's ability to discuss any given writer, historical fact, social, artistic, or religious era. In Lanham's universe, there is no training in cultural memory, and so there is no culture, and no memory beyond the few little bits one might personally accumulate during one's short life and the techniques one needs to use a bus schedule. Lan-

20. The 12th and 13th centuries are marked by laments at the rise of the modern, the eclipse of literature and theology by those who cared only for law and rhetoric [25]. Like Lanham today, rhetoricians of the Middle Ages hawked their courses by promising the basic ability to write notes, make money, and be practical, offering courses "short and practical, with no time wasted on outgrown classical authors but everything fresh and snappy and up-to-date, ready to be applied the same day if need be!" [26].

21. Professor Philip Rieff, referenced here, was Benjamin Franklin Professor at the University of Pennsylvania, and a founder of *Daedalus*. Whether or not they have heard of each other, Rieff and Lanham are fairly thought of as opponents in the authority of text. My citation here is warning of my bent.

ham's "democratic" information systems university has no place to exercise the facilities of aesthetic or moral judgment, and so he would not be surprised, I hope, if his students turned out incapable of exercising such judgments.[22]

There is a great irony in these essays, though. When he is not pushing computer/rhetoric's claim to the throne, no one is more outraged than Lanham at an "historical ignorance of educational history, and adulatory misreading of Plato" (178). Taking on Allan Bloom (in the style of Lilliputians taking on Gulliver), Lanham blasts him for "re-arranging history" (178). There *are,* it seems, limits to the interpretation that Lanham will allow.[23] But how can anyone "misread" Plato? Surely the reader can adjust the writing by becoming the writer; "mix and match" Lanham encourages us (129). Make your own Playdo. Lanham has no ground here for criticism: he despoiled that ground some time, and many pages, ago.

DIAMONDS IN THE ROUGH

Slogging through the endless citations to authors that none of his ideal students will ever read, past the Latin and Greek, past staccato cites to Racine and Ortega y Gasset,[24] beyond the tiresome academic broadsides fired against others in academe, one finds tasty morsels here.

---

22. Lanham shrugs this sort of thing off. Show me, he says, how the standard liberal arts education has made this a more moral world. Have not, he asks rhetorically (sorry), the products of Western civilization slaughtered millions? Have not the genteel citizen descendants of Newton made nuclear bombs? Did not the spy Anthony Blunt who stood as the apotheosis of the well-bred Western civilized man, commit treason (179 et seq.)? Where were the moral claims of classic Western civilization then, eh? This is really a remarkably irritating form of argument, recalling the worst stereotypes of the Greek sophists to whom Lanham traces his intellectual roots. Anecdotal evidence is an oxymoron: Western civilization has also brought us the Geneva Convention, the Golden Rule, cricket, and political democracy. Not bad; but so what. These things are in no balance. There is not enough time left in the life of this universe to weigh the cost of our history with an imagined one.

23. And Lanham is incredulous that Jacques Derrida has apparently not read Kenneth Burke (56). But who *reads*? we ask Lanham.

24. "Surely here we must find what Ortega y Gasset called the 'height of the time,' which ought to animate a liberal education" (254). That's it on that author. Not useful. In another place, Lanham explains his synopsis of E. D. Hirsch's *Cultural Literacy: What Every American Needs To Know* by writing, "This is Dewey out of Rousseau's *Emile* by Wordsworth's *Prelude"* (170).

Computers and Copyright

Lanham points out in a number of his essays that copyright law is challenged by the ubiquity of digitized information. Serious problems arise concerning ownership of rights in continually revised, modified, and amended electronic creations.[25] The blurring of lines between creator and viewer and consumer create difficult issues (18–19, 134).[26] "Electronic information seems to resist ownership," Lanham writes (19). How, he asks, does one protect the "quiddity" of a work when the work is essentially digital and can be manifested in so many different contexts? What do we do when "all the constants that lend substance to the 'substantial similarity' test dissolve in the digital environment" (276)?[27]

This fungibility brings on a deep unease, well and simply captured by John Seabrook. He had written a fun article on email and Microsoft's Bill Gates [33], and in responding to ensuing email from his readers and requests for reprints, he wrote [34]:

> I...reminded people (sweetly) that the copyright belonged to me, and that *The New Yorker* held certain reprint rights, and that the usual procedure was to ask my permission and, in some cases, to pay me. I realize that in saying this I was violating the spirit of the Net, which is all about sharing, and I was sort of surprised at myself, but the idea that my piece could be completely out of my or *The New Yorker's* control, that it could be reproduced thousands or millions of times, that it could be rewritten or reedited or even re-bylined sort of frightened me.

The point goes far beyond the matter of text. The computer and its algorithmic peephole into the universe generate concordances among a wide variety of phenomena: snowflakes and sound, heartbeats and cartoon faces, cancer genes and speech wave form graphs [35,36]. Computing

25. But Lanham is wrong if he thinks that digitized data reveals the problem for the first time. Photographs may infringe a ballet choreography, *Horhan v. Macmillian*, 787 F.2d 157 (2d Cir. 1986), and motion pictures can infringe a book, *Stewart v. Abend*, 495 U.S. 207 (1990).

26. The issues become especially pronounced in what might be considered an extrapolation of interactive multimedia, virtual reality. "In cyberspace, everyone is the author, which means that no one is an author: the distinction upon which it rests, the author distinct from the reader, disappears. Exit author..." [31]. See also references cited at note 30 below.

27. The "substantial similarity" test for copyright infringement is discussed, for example, in *Worth v. Selchow & Richter Co.*, 827 F.2d 569 (9th Cir. 1987). See generally [32] for a discussion of what "substantial" might mean in various contexts.

machines present the view that significant portions of reality can fully be captured by a code, which in turn must rewrite our use of such notions as "analogy," "similarity," and thus "induction." The interest here is not so much the surface morphing (such as of fonts) which so excites Lanham; it is, I suggest, more the possibility of *algorithmic* mutations and parallels which so richly challenge our thinking.

The matters briefly raised by Lanham with respect to digital data are the questions the courts are asking themselves now, as they confront the issues of software copyright—software, which makes no bones about its digital skeleton, and whose skeleton the courts will not hesitate to protect.[28] Lanham raises these issues, and having writ, moves on; one may find discussions in the works of others.[29]

Public and Private Selves

> *So long as it is our habit to confuse art with life, what appears on-stage will appear off; and what appears off-stage will be staged.*
>
> —Philip Rieff [42]

Secondly, more tantalizing and more briefly, Lanham also notes the incipient questions on the relationship between public and private selves in the context of an electronic world (144, 205, 219). Lanham alludes to the development of the public, self-conscious theatrical self, animated by themes of manner and decorum. Later, in another essay, Lanham suggests that computers will allow more play here: more control over the public personas and more room in the electronic and networked world to adopt and enhance a multitude of such personas. That, in turn, can aid the nurture of the private self, created and defined precisely in juxtaposition with public personas.

As long as these lines are respected—the public and private, what is shown and what is not, what is common and what is reserved, what is in context proper and correct and what is not; in a word, the concept of *privacy*—then may the infinite electronic personalities guard and shape the human self.

But this is, or should be, a touchy topic for Lanham. Privacy is defined by fuzzy boundaries; where the environment gradually leaves off,

28. Object code, source code, microcode, and so on are all protected. See, e.g., *Brown Bag Software v. Symantec Corp.*, 960 F.2d 1465 (9th Cir. 1992); *Apple Computer, Inc. v. Franklin Computer Corp.*, 714 F.2d 1240 (3d Cir. 1983), *cert.dis.*, 464 U.S. 1033 (1984). See generally [37].

29. See, e.g., [36, 38–40].

there does the self start. Public convention rules in the electronic world, no less than in its mundane parallel. These conventions, like anything else, are learned. They are learned by the common distribution and examination of authoritative texts. These generate our manners and morals—our shared assumptions—and allow communications. For true-blue communication requires knowing enough about the recipient to know how much, and what, data to send; we get that only through cultural knowledge. All these shape the boundaries; the culture makes character [43].

Read what Lanham has in store for his students and ask whether they will be ready to deal with the modern environment, mundane and electronic, and still sustain a deep sense of self. The risk here is that all our exterior selves—our electronic personas, which act by fax, on the phones, online, and across the network—will be all that there is, as useful as viscera turned inside out, twisting in the wind. This is the risk that "technics is invading and conquering the last enemy—man's inner life, the psyche itself..." [44].

"Private selves are created by public ones," Lanham tells us (220). But not only public ones. There is no privacy there; just endless display. A thousand effervescent selves whispering in the ether are no guarantee of anything inside.

References

[1] Tzonis, Alexander, *Hermes and the Golden Thinking Machine,* Cambridge, MA: MIT Press, 1990, p. 106.

[2] Lanham, Richard A., *The Electronic Word: Democracy, Technology, and the Arts*, Chicago, 1993, p. 5.

[3] Reif, Rita, "Browsing Through the Art of Gutenberg's Galaxy," *The New York Times* p. H32 (Feb. 6, 1994).

[4] Hofstader, Douglas R., "Metafont, Metamathematic, and Metaphysics: Comments on Donald Knuth's Article 'The Concept of a Meta-Font'," reprinted in *Metamagical Themas: Questing for the Essence of Mind and Pattern*, New York: Basic Books, 1985.

[5] Pickover, Clifford A., *Computers and the Imagination,* New York: St. Martins Press, 1991, p. 26.

[6] Beard, Henry, *Latin for Even More Occasions,* Glasgow, 1991.

[7] Russo, Jack, and Jamie Nafziger, "Software 'Look And Feel' Protection in the 1990s," 15 *Hastings Comm/Ent L.J.* 571 (1993).

[8] Derwin, Douglas K., "It's Time To Put 'Look and Feel' Out To Pasture," 15 *Hastings Comm/Ent L.J.* 605 (1993).

[9] Woolley, Benjamin, *Virtual Worlds*, Cambridge, MA: Blackwell, 1992.

[10] "New Dell Consumer Survey Shows Fear of Technology," 8 *TWICE* 37 (Oct. 18, 1993).

[11] Sandler, N., "Panic Gluttons," *Technology Review* 72 (Oct. 1993).

[12] Rheingold, Howard, "PARC Is Back," 2.02 *Wired* 91, 92 (Feb. 1994).

[13] *Wired* 116 (Sept./Oct. 1993).

[14] 1 *Virtual Reality World* (Summer 1993) p. q.
[15] Sheridan, Thomas B., *Telerobotics, Automation, and Human Supervisory Control,* Cambridge, MA: MIT Press, 1992, p. 96.
[16] Ruelle, David, *Chance and Chaos,* Princeton, 1991.
[17] Ruelle, David, *Chance and Chaos,* Princeton, 1991, p. 54.
[18] Gleick, James, *Chaos: Making a New Science,* New York, 1987.
[19] Shakespeare, William, *King Lear,* Baltimore, 1958, pp. 17–18.
[20] Kerr, Clark, *The Uses of the University,* New York, 1963, pp. 41, 75, 86, 94.
[21] Kerr, Clark, *The Uses of the University,* New York, 1963, p. 121.
[22] Smith, Laura B., "High-Wire Act: Balancing Your Network Needs on ATM, ISDN, and Other High-speed Technologies," 11 *PC Week* 17 (Jan. 21, 1994).
[23] Hardison Jr., O. B., "Education for Utopia," in *High Technology & Human Freedom,* Lewis H. Lapham (ed.), 1985, p. 41.
[24] Haskins, Charles Homer, *The Rise of Universities,* Ithaca, 1957.
[25] Haskins, Charles Homer, *The Rise of Universities,* Ithaca, 1957, p. 53.
[26] Haskins, Charles Homer, *The Rise of Universities,* Ithaca, 1957, p. 32.
[27] Kerr, Clark, *The Uses of the University,* New York, 1963, p. 119.
[28] Haskins, Charles Homer, *The Rise of Universities,* Ithaca, 1957, p. 93.
[29] Rieff, Philip, *Fellow Teachers,* New York, 1972, p. 2.
[30] Rieff, Philip, *Fellow Teachers,* New York, 1972, p. 12.
[31] Woolley, Benjamin, *Virtual Worlds,* Cambridge, MA: Blackwell, 1992.
[32] Nimmer, M., and D. Nimmer, *Nimmer on Copyright,* § 13.03 [A], 1992.
[33] Seabrook, John, "Email From Bill," *The New Yorker* (Jan. 10, 1994).
[34] Seabrook, John, "In The Mail," *The New Yorker* 9 (Feb. 7, 1994).
[35] Pickover, Clifford A., *Computers, Pattern, Chaos and Beauty,* New York: St. Martin's, 1990.
[36] Karnow, Curtis, "Data Morphing: Ownership, Copyright, and Creation," in this volume.
[37] "Pragmatism in Software Copyright: *Computer Associates v. Altai,*" Note, 6 *Harvard J. of Law and Technology* 83 (1992).
[38] Davis, G. Gervaise, III, "The Digital Dilemma: Coping With Copyright in a Digital World," 8 *The Computer Law Association Bulletin* 4 (1992).
[39] Rodarmor, William, "Rights of Passage," *NewMedia* 49 (Sept. 1993).
[40] Morgan, Charles, "Sampled Unto Death," *NewMedia* 17 (Sept. 1993).
[41] Samuelson, P., and R. Glushko, "Intellectual Property Rights for Digital Library and Hypertext Publishing Systems," 6 *Harvard J. Law and Technology* 237 (1993).
[42] Rieff, Philip, *Fellow Teachers,* New York, 1972, p. 110.
[43] Rieff, Philip, *Freud: The Mind of the Moralist,* Chicago, 1959, p. 368.
[44] Rieff, Philip, *Freud: The Mind of the Moralist,* Chicago, 1959, p. 256.

Part VIII
Terminus

We must reflect more deeply than we do on the effect of modern technological life upon the emotional and instinctual development of man. It is quite possible that the person whose life is divided between tending a machine and watching TV is sooner or later going to suffer a radical deprivation in his nature and humanity.

—Thomas Merton [1]

Portions of these chapters suggest that the two systems—traditional law and rapidly developing computer technology—are operating at such different speeds that any meeting must be a collision. This suggests that we need a "new" law for the computer or networked world, new rules and procedures for the electroverse. It may also suggest that no law will be effective. In this book, I have turned that stone around and around and provided different answers.

Courts have come out on that issue differently, too. The Washington state Supreme Court wrote that "traditional expectations" were emphatically not affected by technology developments.[1] And it's true that most cases do not present difficult questions. Outright fraud, wholesale copying of an entire text or software, the exposure of trade secrets such as the formulas for Coca-Cola or of source code—in any environment, in any medium, at any speed, regardless of the court or the community evaluating the action—are all obviously illegal. Technology here is a vehicle, a new and surprisingly mutating one perhaps, but just a vehicle; the nature of its cargo is constant regardless of its method of transportation.

1. *Washington v. Faford*, 128 Wash.2d 476, 910 P.2d 447 (Washington Sup.Ct., Feb. 1, 1996).

Other courts have suggested by contrast that there are serious problems with the digital/legal interface, and that new law is needed for the new online world.[2] And that is true, too. What *is* "fraud" in an online meeting place where everyone is an electronic persona and has clothed himself with imaginary attributes? What is wrong with downloading software, at least when one is under the impression that the author posted the program for all to view? And can we tell what a trade secret is when it is found somehow, buried deeply in a distant computer—but discovered and retrieved by an intelligent agent program?

Legal rules, as such, may remain valid; the factual hooks in the world on which the rules depend for their application and enforcement are increasingly nebulous, however. Each of the three legal issues alluded to earlier, for example (fraud, copyright, and trade secrets), rely on borders, or distinctions in the real world between juxtaposed states. Fraud involves deception: appearing to be, or to say, one thing when the contrary is true. Copying can be illegal or permitted: it may depend on whether there is an implied license for the copying. The circumstances are examined to determine whether a "reasonable" person would believe he had permission to copy. Trade secrets are illegally secured when "improper means" are used to obtain them; by contrast, a trade secret loses its protected status when its owner does not adequately hide it from outsider scrutiny. The rules are fairly clear. But their applications rapidly sink into a quagmire when, in the digital networked context, appearances are both true and false—as real as virtual reality; when no one knows the circumstances of a digital posting of software and no one is, indeed, sure of its author (if there was one); or when hiding or finding a "secret" file depends on the sophistication of an intelligent software agent and on the ignorance of a human user of the electronic network.

The interplay of law and advanced computer technology is a conceit, and, like any image, can only be roughly right. These two areas are both vast terrains, operating at multiple levels and through a variety of means and expressions. The result in a given case may depend on a statute or cases decided from one appellate court but not another; or, sometimes, on the judge, or the make-up of the jury; or on tactical decisions made by lawyers. The legal system operates at all these levels. And the terrain of technology is no more consistent: Simply because an action is undertaken by a machine does not *ipso facto* put it in a new and unusual context. Many computer uses, such as writing a book, are entirely routine. Law and technology intersect at many points, and only some of those are at the edge, where the old metaphors seem stripped away.

2. *It's In The Cards, Inc. v. Fuschetto,* 193 Wis.2d 429, 535 N.W. 2d 11, 15-16 (Wis.Ct.App.1995).

Both descriptions are accurate. The classic legal metaphors are sometimes effective, in the sense of being powerful predictors of what a court will do—and sometimes those metaphors are useless.

This dual description is consistent with many episodes in the development of the law. As the legal machinery over time decides cases, experiments with rationales, and understands the technology, surely we will always find—by definition?—a creeping orthodoxy, more routine cases, and always, out ahead in the vanguard, an advancing edge of novelty. The image is of expanding nebulae: the exploding star blows out its turbulent skin of gases, leaving behind an expanding globe of quiescence.

Digital technology is not, of course, the first new technology that has challenged the law. Water power and mills changed the face of American law in the 1800s [2], and the rise of mass-market consumer products brought us modern products liability law and its rebirth of strict liability. A marvelous doctrine known as *ancient lights* used to bar the building of tall structures too close to neighboring property because that would cut off long-established rights to natural light [3]. Urbanization could not stand that doctrine too long. After some nasty, difficult cases, it went the way of the mastodon, to be replaced with rules of eminent domain more suited to growth. In each of these situations, the difficult, twisted, nebulotic expansion of legal doctrine was followed by the tranquillity of routine application. Deep changes in the society shifted the landscape, and the law found other ways to link into the new world.

It is fair to speculate whether digital technology will follow this same avenue. In some areas, it is apparent that it will do so. For example, for some time there was quite a stir on whether domain names, invented terms used as Internet addresses such as *mcdonalds.com* and *disney.com*, were trademarks, and so subject to attack under the federal trademark statute, the Lanham Act. These are *cyberspace addresses*, went one argument, no more brand names than is 1250 Connecticut Avenue, Washington, D.C. But in the navigational nightmare of the World Wide Web, these domain names did in fact act as calls to consumers, directing purchaser attention and suggesting affiliation with various companies—just as do trademarks such as EXXON and XEROX. And so that dispute probably has been laid to rest, moving from the turbulence of new case law to the still waters of established doctrine.[3]

3. The owner of one domain name was barred from using it because the domain name was "confusingly similar" to federally registered trademark. *The Comp Examiner Agency v. Juris Inc.* (No. 96-0213 C.D.Cal. 1996). Domain names are issued by Network Solutions, Inc., and its policy with respect to the role of trademark law as of August 1996 is at http://rs.internic.net/domain-info/internic-domain-6.html. Under that policy, those with prior federally registered trademarks are protected and can block those wishing to use a confusingly similar domain name.

♦ ♦ ♦

But there is a lurking suspicion that, even as the kinks in some problems are worked out, many legal issues, including important ones, will remain intractable. If that suspicion has foundation, it is surely the nature of the computer itself and its networked incarnation.

A vital characteristic of computers its their self-reference; which is to say, the facility of computers—of digital logic—to self-reference at the expense of the real, physical world. Baudrillard has written of [4]

> an unconditional realization of the world by the actualization of all data, the transformation of all our acts all events into pure information: in short, the final solution, the resolution of the world ahead of time by the cloning of reality and the extermination of the real by its double. [4]

"[H]ere technology produces a positive (homeopathic) side-effect. The integrated circuit loops back on itself, ensuring, as it were, the automatic deletion of the world" [6]. The significance of this absolute self-reference is this: It undermines the model, sketched out earlier, of how persisting legal doctrine slowly finds its way into a new technology, such as new markets, the new worlds of transportation, industry, or mass consumer production. It is not just that legal doctrines, such as fraud, copyright, or trade secrets, need more time to find their "hooks" in the real world. Rather, the problem is novel: Legal doctrines in the advanced computer context may not find their hooks in the real world at all, but in imaginary ones. Judges and lawyers are not attempting to bend the legal rules to fit a new reality, but rather to a new hallucination. William Gibson, seer of cyberspace and author of the text *Neuromancer*, calls cyberspace a consensual hallucination. That, the legal system might accommodate. But cyberspace is not an agreed-upon state or reality; it is a soup of inconstant mass hallucinations *which eviscerate the unitary consensual state*. With their referents ever morphing, legal rules are cut adrift.

Information technologies supersede, as physical realities fade. The self-referential nature of computer technology is apparent throughout the chapters in this volume. Artificial intelligence takes the place of acts of God and other uncontrollable "physical" events; public fora such as markets and town squares are swept over with computer conference areas.

4. A wonderful model of an entirely self-referential digital universe used as a model for scientific understanding is found in the title and text of a book on the game of Life in [5].

National law is replaced with the varieties of private fiat. Objective physical property takes second place to the endlessly mutable effervescence of digital properties. "I don't run a trucking company," one trucking company executive said. "I run an information technology organization." As many of these chapters record, the infrastructure is shifting to the electroverse. Concomitantly, technology law will not look so much to interpret and judge actions in the *physical* world, but rather the acts of our varied electronic shadows.

In short, whereas in the past large adjustments in the legal system have resulted in an eventual accommodation of a new social or technological reality, such a process now will be more likely to founder for lack of an emergent unitary, objective, agreed-upon world to which the law's accommodation might be sought.

♦ ♦ ♦

The current legal system developed in the context of an orderly natural world, in which time and space were prominently serial phenomena, and in which there were plain presumptions (or biases) delineating a natural context and background from active (usually human) agents which superimposed themselves on the natural order. Ground and figure, context and efficient agent, nature and human artifact—these juxtapositions defined the realm governed by law. On one side of the divide, the physical world is free of the legal regime. King Canute proved that, when his civil authority failed to move the tides. The truth was that the king and his law had their way, instead, on the other side of the divide: over humans and their artifacts. The peculiar notion of "ownership" is our way of moving things over to the command of the law. Property "ownership" selects parts of the natural and physical world and then transmutes them into extensions of the human world, bringing them under the jurisdiction of justice.

Difficulties arise when *the entire context is artificial*, and the old dichotomies fail. Nothing is beyond the law when we can encode the world and copyright the universe. The wretched irony is that, as the electroverse expands, it does so by replacing physical nature and physical laws with "owned" things such as software, trademarks, patented sets of algorithms, and other denizens of this man-made universe. All these are, by definition, subjects of the legal system. So the law finds itself increasingly called upon, even as its power to adjudicate diminishes.

♦ ♦ ♦

The dawn of the fully artificial context is not simply the transformation of context into a self-image; this is not simply man's transmutation of the natural world into parking lots and power plants, the concretization of the world as it were. We are rather redefining ourselves as technological beings. Instead of anthropomorphizing computers, we are using the puissant metaphor of computers to define ourselves, our own minds. It is not an inherently obvious correlation. Convincing metaphors for the mind have been found in everything from clockworks to wild animals, and of course the digital metaphor was not available before the widespread use of computers. Even now many would argue against the explanatory power of analogizing computers and human minds [7,8].

Nevertheless, many do now think of people as information processors, problem solvers, memory holders. Huberman, for example, writes, "[T]he similarity between distributed social or biological intelligence and its computational counterpart is pervasive, and suggests the possibility of using computers to study general issues of collective problem solving" [9].[5] And we are using the model of the electrosphere to understand, and take the place of, the physical world. "By making a machine think as a man, man recreates himself, defines himself as machine" [12].[6]

In the digital world, man and machine are approaching each other as cybernetic cousins, both fundamentally alike, concentrations of data and algorithms set in a sea of digital data and algorithms. As we spend more and more time with our computers, and increasingly define ourselves through our work and communications with and through computers, we make ourselves machine tractable; in short, computable.

This blurs the human and inhuman. It is hope for some[7] and a horror for others [15]:

> Am I a man or a machine? This anthropological question no longer has an answer.... The new technologies, with their new

---

5. Huberman accurately suspects that his perspective is a function of extant technology: "Just as social structures are being used as inspiration for the design of network systems, so can networks point the way to a deeper understanding of the complex interactions that define natural social systems.... This symbiosis, however, may be a function of the very technology with which we approach and shape our world" [10]. See [11] (brain structure and function expressly described as similar to digital computers).

6. "By the late 20th century, our time, a mythic time, we are all chimeras, theorized and fabricated hybrids of machine and organism; in short, we are cyborgs. The cyborg is our ontology; it gives us our politics. The cyborg is a condensed image of both imagination and material reality, the two joined centers structuring any possibility of historical transformation" [13].

7. Mr. Bolter appears to embrace the conjunction of man and machine in a combination he terms "synthetic intelligence" [14].

> machines, new images and interactive screens, do *not* alienate me. Rather, they form an integrated circuit with me.... Alienation of man by man is a thing of the past: now man is plunged into homeostasis by machines.

It is this hybridization in a digital context which dissolves the distinctions between agencies and their environment, and between humans, artifacts, and their natural context. The old question—*are machines intelligent like us*?—will be answered affirmatively; not because computers are so phenomenally advanced, but because we have made ourselves in *their* image, overcome by the metaphors of fluidity, synthetic synesthesia, the hyperreal, the "lethal illusion of perfection" [16].[8]

This completes the feedback loop in which both humans and their environment increasingly comprise a digital, computer-mediated fantasy world. [9] That is a place where human actors are no more or less responsible than their machine counterparts, and where there are so many agents, human and otherwise, that fixing blame and legal liability is lost in an impossibly impenetrable tangle of complexity. We are left with utter diffusion, the "abandonment of responsibility" [22].

♦ ♦ ♦

In 1897, before his tenure on the United States Supreme Court and while he was with the Massachusetts Supreme Judicial Court, Oliver Wendell Holmes described the law as "systematized prediction"; that is, predictions about what a judge will or will not do in a circumstance. The law is not, said Justice Holmes, a mathematical system that can be worked out in the abstract or derived from rules of morality. It is practical, a series of forecasts about what is likely happen in court [23]. Holmes looked to consensus and social instinct for the justification of the results of this system: "The law can ask no better justification than the deepest instincts of

8. See generally [17]. The metaphor of the hybrid has recently been reified. See [18] (direct connection and simulation between silicon chip and leech nerve cell). See also [19] (reviewing books on cyborgs—an "odd marriage of flesh and technology"—including [20,21]).

9. The day these words were redrafted in November of 1996 I heard of a patent applied for by IBM, known familiarly as the "wet wire" because it turns a human being into a conductive wire, a section of an electronic circuit. A small, credit-card sized instrument carried in, say, a pocket contains electronic information, such as credit card numbers, phone card PINs, or other information. The instrument sends a current a billionth of watt through the body, such that an instrument built with the corresponding receiver will pick up the signal. The receiving instrument may be a phone, or cash machine, or any other digital machine: Without other contact, the mere touch of the human will send digital information to the machine.

man" [24]. Justice Holmes believed, had to believe, that those deepest instincts were of necessity common to humanity.

Practicing lawyers and judges intuitively accept Holmes' characterization, for we are in the business of predicting outcomes. From this practical vantage, we have reason for great concern when the legal system loses its predicative attributes, loosed from the anchor of consensus or "instincts." The threat is the loss of the law.

The threat is not, of course, solely a function of rapidly advancing computer technology. The waxing and waning of consensus is a delicate thing, rising and falling on the tides of politics, social and geographical mobility, education and its decay, the rise of the therapeutic culture and the erosion of authority, and a host of other social shifts. But advanced computer technology does conspire with these shifts; it is entwined with them.

Thus, the challenge for the legal system is to establish, for the nebulous electroverse, standards and foundational consensus similar to those offered by the relatively objective physical world and revealed by physical laws.

These standards are in development. They are a hybrid of technical requirements, computer protocols, and social etiquette. They range from standards for programmers to rules for the interaction of software search robots and their host web sites, standards for digitally encrypted signatures to the Turing Registry and legal recognition for epers proposed in this volume. Members of Internet communities known as MUDs[10] sign onto rules of behavior and manners involving the "rights" of electronic personalities: violations of the rules can lead to expulsion from the digital Eden or lesser punishment including public humiliation and suspension of privileges. In many areas on the Internet, there are generally accepted standards barring personal attacks, advertising, certain subjects, and so on. Transgressors are chastised or erased.

Not surprisingly, much of the work here concerns money and business, and the increasing use of the Internet for commercial purposes has produced renewed efforts on standards. For example, private companies have issued protocols for the use of digital cash; standards have been proposed for the use of trusted intermediaries in electronic commerce [25,26]. Evolving standards for encryption are instructive and reveal the multiple

---

10. As explained by an online frequently asked questions (FAQ) file, a "MUD (Multiple User Dimension, Multiple User Dungeon, or Multiple User Dialogue) is a computer program which users can log into and explore. Each user takes control of a computerized persona/avatar/incarnation/character. You can walk around, chat with other characters, explore dangerous monster-infested areas, solve puzzles, and even create your very own rooms, descriptions and items." MUDs are further discussed in Chapter 10 of this volume.

levels at which consensus may be forming. Users, industry, and governments are examining potential agreements on issues from low-level specific sets of algorithms to security levels (or encryption "strengths"), which will be generally acceptable for export and international use, to the rules governing employers' rights to decrypt employees' email, and conversely the limits of employees' email privacy in the workplace. At the time of this writing, a draft of a new section of the Uniform Commercial Code is circulating, which may harmonize the law on the general effect of software licenses, including Internet negotiations and distribution.

But, finally, the issue is not whether some consensus is forming within the high-technology and computer industries. The issue is, more specifically, the extent to which there are consensus and standards sufficiently ubiquitous, durable, and obvious that the legal system can cite them. Nothing else can provide the requisite foundation that Justice Holmes referred to as "social instinct."

If the law must place hooks into the digital world it seeks to regulate, then one of two processes must occur. Either the digital electroverse slows down sufficiently to allow the broad dissemination and percolation of standards and consensus—or else the legal system must develop to meet the challenge.

Perhaps there is a chance for both improvements.

The legal system must surely change. The courts must become more responsive, and the process of dispute resolution must become faster, and less expensive. The legal system must become *reasonable*, with judges as sensitive to the nightmarish expenses and delays of litigation as they are to the mandates of their superior courts and the Constitution. It is a fine court system that honors every punctilio of perfect process, but not if it can be stomached by only a small fraction of the population. Such changes in the legal process are needed not just to accommodate accelerating technology, but for the more general reasons of ensuring access to justice and the civic enfranchisement that entails.

I had thought that it would be safe to assume that the pace of technological change could not be retarded, even if we wished it, and I had thought that it would be odd to wish it. I have generally tracked the new and fast-developing technologies; like a frog's eyes, ours are drawn to motion and change. That which is constant vanishes. But the future is not always shocking. Tempering enthusiasm for the novel, it is possible to see that many of the technologies may find some stability, especially as they move into the consumer markets, where standards, interoperability, consistency, reliability, and simplicity of use tend to dominate. There may be room, there, for a slow seeping of computer technologies. And in that insidious movement, the technology may becomes ubiquitous or nearly so, at least within the industrialized nations and their confréres. This would enable the percolation I mentioned above—percolation that permits the population to gradually form consensus on the uses and effects of technologies. It is difficult to imagine, now, a mundane digital technology that develops at this human speed. But the alternative is one that moves at the speed of light, blinding us, and sundering us from the past and its incarnation in the law.

References

[1] Merton,Thomas, *Life and Holiness,* Image Books, 1963, p. 25.
[2] Horwitz, Morton J., *The Transformation of American Law: 1780–1860,* Cambridge, MA: Harvard University Press, 1977.
[3] Horwitz, Morton J., *The Transformation of American Law: 1780–1860,* Cambridge, MA: Harvard University Press, 1977, p. 44.
[4] Baudrillard, Jean, *The Perfect Crime,* Verso, 1996, p.25.
[5] Poundstone, William, *The Recursive Universe: Cosmic Complexity and the Limits of Scientific Knowledge,* Chicago, 1985.
[6] Baudrillard, Jean, *The Perfect Crime,* Verso, 1996, p. 41.
[7] Crombag, Hans F. M., "On the Artificiality of Artificial Intelligence," 2 *Artificial Intelligence and the Law* 39, 41 (1994).
[8] Searle, John R., *The Rediscovery of the Mind,* Cambridge, MA: MIT Press, 1992.
[9] Huberman, Bernardo A., "Towards a Social Mind," 2 *Stanford Human Rev.* 103, 106 (1992).
[10] Huberman, Bernardo A., "Towards a Social Mind," 2 *Stanford Human Rev.* 109 (1992).
[11] Dawkins, Richard, *The Selfish Gene,* Oxford, pp. 48–49, 276.
[12] Bolter, J. David, *Turing's Man,* Univ. North Carolina, 1984, p. 13.
[13] Haraway, Donna J., "A Cyborg Manifesto: Science, Technology and Socialist-Feminism in the Late Twentieth Century," in *Simians, Cyborgs and Women; The Reinvention of Nature,* New York: Routledge, 1991, p. 150; quoted in Margaret Morse, "What Do Cyborgs Eat? Oral Logic in an Information Society," in *Cultures on the Brink: Ideologies of Technology,* Gretchen Bender and Timothy Druckrey (eds.), Seattle: Bay Press, 1994, pp. 157, 198.
[14] Bolter, J. David, *Turing's Man,* Univ. North Carolina, 1984, p. 238.
[15] Baudrillard, Jean, *The Transparency of Evil,* Verso, 1993, pp. 57–58, 59.
[16] Baudrillard, Jean, *The Illusion of the End,* Stanford, 1994, p. 101.
[17] Baudrillard, Jean, *The Perfect Crime,* Verso: 1996.
[18] Browne, Malcolm W., "Neuron talks to Chip and Chip to Nerve Cell," *The New York Times* p. B7, col. 4 (Aug. 22, 1995).
[19] Rothstein, Edward, "Connections," *The New York Times* p. C3 (Jan. 8, 1996).
[20] Gray, Chris (ed.), *The Cyborg Handbook,* 1995.
[21] Mazlish, Bruce, *The Fourth Discontinuity: The Co-Evolution of Humans and Machines,* Yale, 1995.
[22] Sheridan, Thomas B., *Telerobotics, Automation, and Human Supervisory Control,* Cambridge, MA: MIT Press, 1992, p. 241.
[23] Holmes, O. W., "The Path of the Law," X *Harvard Law Review* 61 (1897).
[24] Holmes, O. W., "The Path of the Law," X *Harvard Law Review* 70 (1897).
[25] Froomkin, A. Michael, "The Essential Role of Trusted Third Parties in Electronic Commerce," 75 *Oregon Law Review* 49 (Spring 1996).
[26] Stefik, Mark, "Trusted Systems," *Scientific American* 78–81 (March 1997).

About the Author

Curtis E. A. Karnow is a partner in the San Francisco offices of the nationwide law firm Sonnenschein Nath & Rosenthal. He grew up in North Africa, France, and Hong Kong, and is a graduate of Harvard College and the University of Pennsylvania law school where he was an editor of the Law Review. A former federal prosecutor and currently occassional judge *pro tem* for state trial courts in the San Francisco area, Mr. Karnow specializes in intellectual property litigation, and high technology and computer law. His clients have included a worldwide telecommunications company; an Internet search engine company; software developers, distributors, and users; makers of Internet tools and applications, encryption software, and numerous global computer and home video game manufacturers and publishers. He is the author of many papers on computer law, litigation, and arbitration, and he has spoken in a variety of business, legal, and university fora throughout the United States and abroad. Mr. Karnow serves as editorial advisor to *Leonardo*, and to the University of California's (Hastings) journal of communications and technology law, *Comm/ENT*. He lives in Marin County, California with his wife and two young children, all of whom stoically suffer his interests in technology. Mr. Karnow can be reached through the Internet at cek@sonnenschein.com.

Index

The Artech House Telecommunications Library

Vinton G. Cerf, Series Editor

Access Networks: Technology and V5 Interfacing, Alex Gillespie

Advanced High-Frequency Radio Communications, Eric E. Johnson, Robert I. Desourdis, Jr., et al.

Advanced Technology for Road Transport: IVHS and ATT, Ian Catling, editor

Advances in Computer Systems Security, Vol. 3, Rein Turn, editor

Advances in Telecommunications Networks, William S. Lee and Derrick C. Brown

Advances in Transport Network Technologies: Photonics Networks, ATM, and SDH, Ken-ichi Sato

An Introduction to International Telecommunications Law, Charles H. Kennedy and M. Veronica Pastor

Asynchronous Transfer Mode Networks: Performance Issues, Second Edition, Raif O. Onvural

ATM Switches, Edwin R. Coover

ATM Switching Systems, Thomas M. Chen and Stephen S. Liu

Broadband: Business Services, Technologies, and Strategic Impact, David Wright

Broadband Network Analysis and Design, Daniel Minoli

Broadband Telecommunications Technology, Byeong Lee, Minho Kang, and Jonghee Lee

Cellular Mobile Systems Engineering, Saleh Faruque

Cellular Radio: Analog and Digital Systems, Asha Mehrotra

Cellular Radio: Performance Engineering, Asha Mehrotra

Cellular Radio Systems, D. M. Balston and R. C. V. Macario, editors

CDMA for Wireless Personal Communications, Ramjee Prasad

Client/Server Computing: Architecture, Applications, and Distributed Systems Management, Bruce Elbert and Bobby Martyna

Communication and Computing for Distributed Multimedia Systems, Guojun Lu

Community Networks: Lessons from Blacksburg, Virginia, Andrew Cohill and Andrea Kavanaugh, editors

Computer Networks: Architecture, Protocols, and Software, John Y. Hsu

Computer Mediated Communications: Multimedia Applications, Rob Walters

Computer Telephone Integration, Rob Walters

Convolutional Coding: Fundamentals and Applications, Charles Lee

Corporate Networks: The Strategic Use of Telecommunications, Thomas Valovic

Digital Beamforming in Wireless Communications, John Litva, Titus Kwok-Yeung Lo

Digital Cellular Radio, George Calhoun

Digital Hardware Testing: Transistor-Level Fault Modeling and Testing, Rochit Rajsuman, editor

Digital Switching Control Architectures, Giuseppe Fantauzzi

Digital Video Communications, Martyn J. Riley and Iain E. G. Richardson

Distributed Multimedia Through Broadband Communications Services, Daniel Minoli and Robert Keinath

Distance Learning Technology and Applications, Daniel Minoli

EDI Security, Control, and Audit, Albert J. Marcella and Sally Chen

Electronic Mail, Jacob Palme

Enterprise Networking: Fractional T1 to SONET, Frame Relay to BISDN, Daniel Minoli

Expert Systems Applications in Integrated Network Management, E. C. Ericson, L. T. Ericson, and D. Minoli, editors

FAX: Digital Facsimile Technology and Applications, Second Edition, Dennis Bodson, Kenneth McConnell, and Richard Schaphorst

FDDI and FDDI-II: Architecture, Protocols, and Performance, Bernhard Albert and Anura P. Jayasumana

Fiber Network Service Survivability, Tsong-Ho Wu

Future Codes: Essays in Advanced Computer Technology and the Law, Curtis E. A. Karnow

A Guide to the TCP/IP Protocol Suite, Floyd Wilder

Implementing EDI, Mike Hendry

Implementing X.400 and X.500: The PP and QUIPU Systems, Steve Kille

Inbound Call Centers: Design, Implementation, and Management, Robert A. Gable

Information Superhighways Revisited: The Economics of Multimedia, Bruce Egan

Integrated Broadband Networks, Amit Bhargava

International Telecommunications Management, Bruce R. Elbert

International Telecommunication Standards Organizations, Andrew Macpherson

Internetworking LANs: Operation, Design, and Management, Robert Davidson and Nathan Muller

Introduction to Document Image Processing Techniques, Ronald G. Matteson

Introduction to Error-Correcting Codes, Michael Purser

An Introduction to GSM, Siegmund Redl, Matthias K. Weber, Malcom W. Oliphant

Introduction to Radio Propagation for Fixed and Mobile Communications, John Doble

Introduction to Satellite Communication, Bruce R. Elbert

Introduction to T1/T3 Networking, Regis J. (Bud) Bates

Introduction to Telephones and Telephone Systems, Second Edition, A. Michael Noll

Introduction to X.400, Cemil Betanov

LAN, ATM, and LAN Emulation Technologies, Daniel Minoli and Anthony Alles

Land-Mobile Radio System Engineering, Garry C. Hess

LAN/WAN Optimization Techniques, Harrell Van Norman

LANs to WANs: Network Management in the 1990s, Nathan J. Muller and Robert P. Davidson

Minimum Risk Strategy for Acquiring Communications Equipment and Services, Nathan J. Muller

Mobile Antenna Systems Handbook, Kyohei Fujimoto and J.R. James, editors

Mobile Communications in the U.S. and Europe: Regulation, Technology, and Markets, Michael Paetsch

Mobile Data Communications Systems, Peter Wong and David Britland

Mobile Information Systems, John Walker

Networking Strategies for Information Technology, Bruce Elbert

Packet Switching Evolution from Narrowband to Broadband ISDN, M. Smouts

Packet Video: Modeling and Signal Processing, Naohisa Ohta

Personal Communication Networks: Practical Implementation, Alan Hadden

Personal Communication Systems and Technologies, John Gardiner and Barry West, editors

Practical Computer Network Security, Mike Hendry

Principles of Secure Communication Systems, Second Edition, Don J. Torrieri

Principles of Signaling for Cell Relay and Frame Relay, Daniel Minoli and George Dobrowski

Principles of Signals and Systems: Deterministic Signals, B. Picinbono

Private Telecommunication Networks, Bruce Elbert

Radio-Relay Systems, Anton A. Huurdeman

RF and Microwave Circuit Design for Wireless Communications, Lawrence E. Larson

The Satellite Communication Applications Handbook, Bruce R. Elbert

Secure Data Networking, Michael Purser

Service Management in Computing and Telecommunications, Richard Hallows

Smart Cards, José Manuel Otón and José Luis Zoreda

Smart Highways, Smart Cars, Richard Whelan

Super-High-Definition Images: Beyond HDTV, Naohisa Ohta, Sadayasu Ono, and Tomonori Aoyama

Television Technology: Fundamentals and Future Prospects, A. Michael Noll

Telecommunications Technology Handbook, Daniel Minoli

Telecommuting, Osman Eldib and Daniel Minoli

Telemetry Systems Design, Frank Carden

Teletraffic Technologies in ATM Networks, Hiroshi Saito

Toll-Free Services: A Complete Guide to Design, Implementation, and Management, Robert A. Gable

Transmission Networking: SONET and the SDH, Mike Sexton and Andy Reid

Troposcatter Radio Links, G. Roda

Understanding Emerging Network Services, Pricing, and Regulation, Leo A. Wrobel and Eddie M. Pope

Understanding GPS: Principles and Applications, Elliot D. Kaplan, editor

Understanding Networking Technology: Concepts, Terms and Trends, Mark Norris

UNIX Internetworking, Second edition, Uday O. Pabrai

Videoconferencing and Videotelephony: Technology and Standards, Richard Schaphorst

Wireless Access and the Local Telephone Network, George Calhoun

Wireless Communications in Developing Countries: Cellular and Satellite Systems, Rachael E. Schwartz

Wireless Communications for Intelligent Transportation Systems, Scott D. Elliot and Daniel J. Dailey

Wireless Data Networking, Nathan J. Muller

Wireless LAN Systems, A. Santamaría and F. J. López-Hernández

Wireless: The Revolution in Personal Telecommunications, Ira Brodsky

Writing Disaster Recovery Plans for Telecommunications Networks and LANs, Leo A. Wrobel

X Window System User's Guide, Uday O. Pabrai

For further information on these and other Artech House titles, contact:

Artech House
685 Canton Street
Norwood, MA 02062
617-769-9750
Fax: 617-769-6334
Telex: 951-659
email: artech@artech-house.com

Artech House
Portland House, Stag Place
London SW1E 5XA England
+44 (0) 171-973-8077
Fax: +44 (0) 171-630-0166
Telex: 951-659
email: artech-uk@artech-house.com

WWW: http://www.artech-house.com